Data Analytics

Series editors

Longbing Cao, Advanced Analytics Institute, University of Technology, Sydney, Broadway, NSW, Australia
Philip S. Yu, University of Illinois at Chicago, Chicago, IL, USA

Aims and Goals:

Building and promoting the field of data science and analytics in terms of publishing work on theoretical foundations, algorithms and models, evaluation and experiments, applications and systems, case studies, and applied analytics in specific domains or on specific issues.

Specific Topics:

This series encourages proposals on cutting-edge science, technology and best practices in the following topics (but not limited to):

Data analytics, data science, knowledge discovery, machine learning, big data, statistical and mathematical methods for data and applied analytics,

New scientific findings and progress ranging from data capture, creation, storage, search, sharing, analysis, and visualization,

Integration methods, best practices and typical examples across heterogeneous, interdependent complex resources and modals for real-time decision-making, collaboration, and value creation.

More information about this series at http://www.springer.com/series/15063

Chuan Shi · Philip S. Yu

Heterogeneous Information Network Analysis and Applications

Springer

Chuan Shi
Beijing University of Posts and
   Telecommunications
Beijing
China

Philip S. Yu
University of Illinois at Chicago
Chicago, IL
USA

Data Analytics
ISBN 978-3-319-85855-5   ISBN 978-3-319-56212-4   (eBook)
DOI 10.1007/978-3-319-56212-4

Printed on acid-free paper

This Springer imprint is published by Springer Nature
The registered company is Springer International Publishing AG
The registered company address is: Gewerbestrasse 11, 6330 Cham, Switzerland

Preface

The interacting and multi-typed components in the real-world environment constitute interconnected networks, which can be called information networks. These ubiquitous information networks form a critical component of modern information infrastructure. In recent years, the information network analysis has gained extremely wide attentions from researchers in many disciplines, such as computer science, social science, physics. Particularly, the information network analysis has become a mainstream direction in data mining, database and information retrieval fields in the past decades. The basic paradigm is to mine hidden patterns through mining linkage relations from networked data. The information network analysis is also related to the works in social network analysis, link mining, graph mining and network science.

Contemporary information network analyses are usually based on homogeneous information networks, where there is only one type of objects or links in the network. An example is the author collaboration network which only contains the author object and the co-author relation. These homogeneous information networks usually are the simplification of real interacting systems by simply ignoring the heterogeneity of objects and links or only considering one type of links among one type of objects. However, most real interacting systems contain multi-typed interacting components which can be modeled as heterogeneous information networks which include different types of objects and links. For example, the bibliographic database, like DBLP, can be organized as a heterogeneous information network which includes multiple types of objects (e.g., papers, authors, and venues) and links (e.g., written by/writing relations between papers and authors, published/publishing relations between papers and venues). Obviously, the author collaboration network is implicitly contained in the heterogeneous information network, which can be derived from the written by/writing relation between papers and authors.

Compared to homogeneous information network, the heterogeneous information network can effectively fuse more information and contain richer semantics in objects and links, and thus it forms a new development of data mining. Since the concept of heterogeneous information network is first proposed in 2009, it rapidly became a hot research topic in data mining, and many innovative data mining tasks have been exploited in this kind of networks. In addition, some unique analysis techniques (e.g., meta-path-based mining) are developed to demonstrate the benefits of heterogeneous information networks. Particularly, with the arrival of the era of big data, heterogeneous information networks offer the potential to be an effective way to model and analyze complex objects and their relations in big data.

This book first provides a comprehensive survey of current developments of heterogeneous information network analysis, as well as some novel data mining tasks in this field. This book includes two parts. In the first part, it deeply and comprehensively summarizes the newest developments of this field in Chaps. 1, 2, and 9. This book introduces in-depth understanding of heterogeneous information network in Chap. 1 and investigates the research developments in most data mining tasks in Chap. 2. Furthermore, based on the newest developments and trends, we point out the future research directions in Chap. 9. In the second part, it illustrates the traits of heterogeneous information network analysis through several data mining tasks in Chaps. 3–8. This book presents relevance measure in Chap. 3, ranking and clustering in Chap. 4, recommendation in Chap. 5, fusion learning in Chap. 6, and schema-rich heterogeneous network mining in Chap. 7. Moreover, some interesting prototype systems are discussed in Chap. 8.

The readers of this book are engineers and researchers in the field of data mining, especially social network analysis. It is also suitable for engineers and researchers in artificial intelligences and informatics. More broadly, readers also include those who are interesting in social network analysis in other disciplines, such as statistics, social sciences, physical, and biology. This book can be used in those courses, such as data mining, social network analysis, complex network, advanced artificial intelligences. These courses are suitable for advanced undergraduates or graduate students specializing in computer sciences and related fields. The readers are suggested to quickly understand this field through the first part and deeply study data mining tasks with the second part.

We would like to express our sincere thanks to all those who work with us on this project. First of all, we appreciate Dr. Jiawei Zhang for his contribution in Chap. 6, which makes this book more integrated. Then, we are grateful to our co-authors in the work of heterogeneous information network. They are Xiangnan Kong, Yizhou Sun, Bin Wu, Yitong Li, Zhiqiang Zhang, Jian Liu, Ran Wang, Yuyan Zheng, Jing Zheng, Xiaohuan Cao, Jiawei Hu, Xiaofeng Meng, Chong Zhou, et al. We also wish to thank supporters during writing this book. They are Xin Wan, Xiaoji Chen, Yugang Ji, Houye Ji, Yiding Zhang, Yang Xiao, Binbin Hu, Xiaotian Han, Pudi Chen, Li Song, Govardhana K., Melissa Fearon, Jennifer Malat, et al. In addition, this work is supported by the National Key Basic Research and Department (973) Program of China (No. 2013CB329600), the National Natural Science Foundation of China (No. 61375058 and 61672313), and US National Science Foundation through grant III-1526499. We also thank the supports of these grants. Finally, we thank our families for their wholehearted support throughout this project.

Contents

Chapter 1
Introduction

Abstract In this chapter, we introduce some basic concepts and definitions in heterogeneous information network and compare the heterogeneous information network with other related concepts. Then, we give some popular examples in this field. In the end, we analyze the reason why mining heterogeneous information network is a new paradigm.

1.1 Basic Concepts and Definitions

As we know, most real systems usually consist of a large number of interacting, multi-typed components, such as human social activities, communication and computer systems, and biological networks. In such systems, the interacting components constitute interconnected networks, which can be called information networks without loss of generality. Clearly, information networks are ubiquitous and form a critical component of modern information infrastructure. The information network analysis has gained extremely wide attentions from researchers in many disciplines, such as computer science, social science, and physics. Particularly, the information network analysis has become a hot research topic in the fields of data mining and information retrieval in the preceding decades. The basic paradigm is to mine hidden patterns through mining link relations from networked data. The analysis of information network is related to the works in link mining and analysis [3, 4, 6], social network analysis [20, 34], hypertext and web mining [1], network science [12], as well as graph mining [2].

An information network represents an abstraction of the real world, focusing on the objects and the interactions among these objects. Formally, we define an information network as follows.

Definition 1.1 (*Information network* [27, 28]). An information network is defined as a directed graph $G = (V, E)$ with an object type mapping function $\varphi : V \rightarrow A$ and a link type mapping function $\psi : E \rightarrow R$. Each object $v \in V$ belongs to one particular object type in the object type set A: $\varphi(v) \in A$, and each link $e \in E$ belongs to a particular relation type in the relation type set R: $\psi(e) \in R$. If two links belong

© Springer International Publishing AG 2017

C. Shi and P.S. Yu, *Heterogeneous Information Network Analysis and Applications*, Data Analytics, DOI 10.1007/978-3-319-56212-4_1

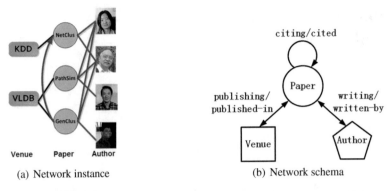

(a) Network instance   (b) Network schema

Fig. 1.1 An example of heterogeneous information network on bibliographic data [27]

to the same relation type, the two links share the same starting object type as well as the ending object type.

Different from the traditional network definition, we explicitly distinguish object types and relation types in an information network and propose the concepts of heterogeneous/homogeneous information network. For simplicity, we also call heterogeneous information network as heterogeneous network or HIN in this book.

Definition 1.2 (*Heterogeneous/Homogeneous information network*). The information network is called **heterogeneous information network** if the types of objects $|A| > 1$ or the types of relations $|R| > 1$; otherwise, it is a **homogeneous information network**.

Example 1.1 Figure 1.1 shows an HIN example on bibliographic data [27]. A bibliographic information network, such as the bibliographic network involving computer science researchers derived from DBLP,[1] is a typical heterogeneous network containing three types of information entities: papers, venues, and authors. For each paper, it has links to a set of authors, and a venue, and these links belong to a set of link types.

In order to understand the object types and link types better in a complex heterogeneous information network, it is necessary to provide the meta-level (i.e., schema-level) description of the network. Therefore, the concept of network schema is proposed to describe the metastructure of a network.

Definition 1.3 (*Network schema* [27, 28]). The network schema, denoted as $T_G = (A, R)$, is a metatemplate for an information network $G = (V, E)$ with the object type mapping $\varphi : V \to A$ and the link type mapping $\psi : E \to R$, which is a directed graph defined over object types A, with edges as relations from R.

---

[1]http://dblp.uni-trier.de/.

The network schema of a heterogeneous information network specifies type constraints on the sets of objects and relationships among the objects. These constraints make a heterogeneous information network semi-structured, guiding the semantics explorations of the network. An information network following a network schema is called a **network instance** of the network schema. For a link type R connecting object type S to object type T, i.e., $S \xrightarrow{R} T$, S and T are the **source object type** and **target object type** of link type R, which can be denoted as $R.S$ and $R.T$, respectively. The inverse relation R^{-1} holds naturally for $T \xrightarrow{R^{-1}} S$. Generally, R is not equal to R^{-1}, unless R is symmetric.

Example 1.2 As described above, Fig. 1.1a demonstrates the real objects and their connections on bibliographic data. Figure 1.1b illustrates its network schema which describes the object types and their relations in the HIN. Moreover, Fig. 1.1a is a network instance of the network schema Fig. 1.1b. In this example, it contains objects from three types of objects: papers (P), authors (A), and venues (V). There are links connecting different types of objects. The link types are defined by the relations between two object types. For example, links existing between authors and papers denote the writing or written-by relations, while those between venues and papers denote the publishing or published-in relations.

Different from homogeneous networks, two objects in a heterogeneous network can be connected via different paths and these paths have different physical meanings. These paths can be categorized as meta paths as follows.

Definition 1.4 (*Meta path* [29]). A meta path P is a path defined on a schema $S = (A, R)$, and is denoted in the form of $A_1 \xrightarrow{R_1} A_2 \xrightarrow{R_2} \ldots \xrightarrow{R_l} A_{l+1}$, which defines a composite relation $R = R_1 \circ R_2 \circ \cdots \circ R_l$ between objects $A_1, A_2, \cdots, A_{l+1}$, where $\circ$ denotes the composition operator on relations.

For simplicity, we can also use object types to denote the meta path if there are no multiple relation types between the same pair of object types: $P = (A_1 A_2 \cdots A_{l+1})$. For example, in Fig. 1.1a, the relation, authors publishing papers in conferences, can be described using the length-2 meta path $A \xrightarrow{writting} P \xrightarrow{written-by} A$, or APA for short. We say a concrete path $p = (a_1 a_2 \cdots a_{l+1})$ between objects a_1 and a_{l+1} in network G is a **path instance** of the relevance path P, if for each a_i, $\phi(a_i) = A_i$ and each link $e_i = \langle a_i, a_{i+1} \rangle$ belongs to the relation R_i in P. It can be denoted as $p \in P$. A meta path P is a **symmetric path**, when the relation R defined by it is symmetric (i.e., P is equal to P^{-1}), such as APA and $APVPA$. Two meta paths $P_1 = (A_1 A_2 \cdots A_l)$ and $P_2 = (B_1 B_2 \cdots B_k)$ are **concatenable** if and only if A_l is equal to B_1, and the concatenated path is written as $P = (P_1 P_2)$, which equals to $(A_1 A_2 \cdots A_l B_2 \ldots B_k)$. A simple concatenable example is that AP and PA can be concatenated to the path APA.

Example 1.3 Consider the examples shown in Fig. 1.2, authors can be connected via meta paths such as "Author-Paper-Author" (APA) path, "Author-Paper-Venue-Paper-Author" ($APVPA$) path, and so on. Moreover, Table 1.1 shows path instances

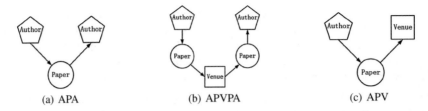

(a) APA   (b) APVPA   (c) APV

Fig. 1.2 Examples of meta paths in heterogeneous information network on bibliographic data

Table 1.1 Meta path examples and their physical meanings on bibliographic data

Path instance	Meta path	Physical meaning
Sun-NetClus-Han Sun-PathSim-Yu	Author-Paper-Author (*APA*)	Authors collaborate on the same paper
Sun-PathSim-VLDB-PathSim-Han Sun-PathSim-VLDB-GenClus-Aggarwal	Author-Paper-Venue-Paper-Author (*APVPA*)	Authors publish papers on the same venue
Sun-NetClus-KDD Sun-PathSim-VLDB	Author-Paper-Venue (*APV*)	Authors publish papers at a venue

and semantics of these meta paths. It is obvious that semantics underneath these paths are different. The APA path means authors collaborating on the same papers (i.e., co-author relation), while $APVPA$ path means authors publishing papers on the same venue. The meta paths can also connect different types of objects. For example, the authors and venues can be connected with the APV path, which means authors publishing papers on venues.

The rich semantics of meta paths is an important characteristic of HIN. Based on different meta paths, objects have different connection relations with diverse path semantics, which may have an effect on many data mining tasks. For example, the similarity scores among authors evaluated based on different meta paths are different [29]. Under the APA path, the authors who co-publish papers will be more similar, while the authors who publish papers on the same venues will be more similar under the $APVPA$ path. Another example is the importance evaluation of objects [13]. The importance of authors under APA path has a bias on the authors who write many multiauthor papers, while the importance of authors under $APVPA$ path emphasizes the authors who publish many papers on those productive conferences. As a unique characteristic and effective semantic capturing tool, meta path has been widely used in many data mining tasks in HIN, such as similarity measure [22, 29], clustering [30], and classification [10].

1.2 Comparisons with Related Concepts

With the boom of social network analysis, all kinds of networked data have emerged, and numbers of concepts to model networked data have been proposed. These concepts have similar meanings, as well as subtle differences. For example, the multitype relational data proposed by Long et al. [18] is an HIN in deed, and the multiview data [15] can also be organized as an HIN. Here, we compare the heterogeneous network concept with those most related concepts.

Heterogeneous network versus homogeneous network. Heterogeneous networks include different types of nodes or links, while homogeneous networks only have one type of objects and links. Homogeneous networks can be considered as a special case of heterogeneous networks. Moreover, a heterogeneous network can be converted into a homogeneous network through network projection or ignoring object heterogeneity, while it will make significant information loss. Traditional link mining [11, 14, 32] is usually based on the homogeneous network, and many analysis techniques on homogeneous network cannot be directly applied to heterogeneous network.

Heterogeneous network versus multirelational network [36]. Different from heterogeneous network, multirelational network has only one type of objects, but more than one kind of relationship between objects. So multirelational network can be seen as a special case of heterogeneous network.

Heterogeneous network versus multidimensional/mode network [31]. Tang et al. [31] proposed the multidimensional/mode network concept, which has the same meaning with multirelational network. That is, the network has only one type of objects and more than one kind of relationship between objects. So multidimensional/mode network is also a special case of heterogeneous network.

Heterogeneous network versus composite network [39, 40]. Qiang Yang et al. proposed the composite network concept [39, 40], where users in networks have various relationships, exhibit different behaviors in each individual network or subnetwork, and share some common latent interests across networks at the same time. So composite network is, in fact, a multirelational network, a special case of heterogeneous network.

Heterogeneous network versus complex network. A complex network is a network with non-trivial topological features and patterns of connection between its elements that are neither purely regular nor purely random [7]. Such non-trivial topological features include a heavy tail in the degree distribution, a high clustering coefficient, community structure, and hierarchical structure. The studies of complex networks have brought together researchers from many areas, including mathematics, physics, biology, computer science, sociology, and others. The studies show that many real networks are complex networks, such as social networks, information networks, technological networks, and biological networks [19]. So we can say that many real heterogeneous networks are complex networks. However, the studies on complex networks usually focus on the structures, functions, and features of networks.

1.3   Example Datasets of Heterogeneous Information Networks

Intuitively, most real systems include multityped interacting objects. For example, a social media website (e.g., Facebook) contains a set of object types, such as users, posts, and tags, and a health care system contains doctors, patients, diseases, and devices. Generally speaking, these interacting systems can all be modeled as heterogeneous information networks. Concretely, this kind of networks can be constructed from the following three types of data.

1. **Structured data**. Structured data stored in database table is organized with entity-relation model. The different-typed entities and their relations naturally construct information networks. For example, the bibliographic data (see the above example) is widely used as heterogeneous information network.
2. **Semi-structured data**. Semi-structured data is usually stored with XML format. The attributes in XML can be considered as object types, and the object instances can be determined by analyzing the contents of attributes. The connections among attributes construct object relations.
3. **Non-structured data**. For non-structured data, heterogeneous information networks can also be constructed by objects and relationship extraction. For example, for text data, entity recognition and relation extraction can form the objects and links of HIN.

Although heterogeneous information networks are ubiquitous, there are not many standard datasets for study, since these heterogeneous information usually exist in different data sources. Here, we summarize some widely used heterogeneous networks in literatures.

Multirelational network with single-typed object. Traditional multirelational network is a kind of HINs, where there is one type of object and several types of relations among objects. This kind of networks widely exists in social websites, such as Facebook and Xiaonei [40]. Figure 1.3a shows the network schema of such a network [40], where users can be extensively connected with each other through connections, such as recording, browsing, chatting, and sending friends applications.

Bipartite network. As a typical HIN, bipartite network is widely used to construct interactions among two types of objects, such as user–item [5] and document–word [16]. Figure 1.3b shows the schema of a bipartite network connecting documents and words [16]. As an extension of bipartite graphs, k-partite graphs [17] contain multiple types of objects where links exist among adjacent object types. The bipartite network has been well studied for a long time. As the simplest HIN, we will not discuss this type of network in this book.

Star-schema network. Star-schema network is the most popular HIN in this field. In the database table, a target object and its attribute objects naturally construct an HIN, where the target object, as the hub node, connects different attribute objects. As an example shown in Fig. 1.3c, a bibliographic information network is a typical star-schema heterogeneous network [22, 29], containing different objects (e.g., paper,

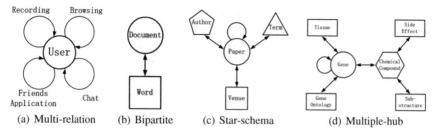

Fig. 1.3   Network schema of heterogeneous information networks

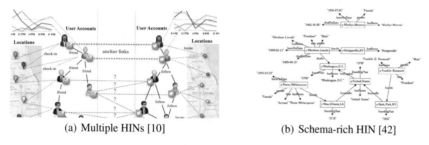

Fig. 1.4   Two examples of complex heterogeneous information network

venue, author, and term) and links among them. Many other datasets can also be represented as star-schema networks, such as the movie data [23, 37] from the Internet Movie Database [2] (IMDB) and the patent data [41] from US patents data.[3]

Multiple-hub network. Beyond star schema, some networks have more complex structures, which involve multiple-hub objects. This kind of networks widely exists in bioinformatics data [8, 33]. A bioinformatics example is shown in Fig. 1.3d, includes two hubs: gene and chemical compound. Another example can be found in the Douban dataset [4] [24].

Besides these widely used networks, many real systems can also be constructed as more complex heterogeneous networks. In some real applications, users may exist in multiple social networks, and each social network can be modeled as an HIN. Figure 1.4a shows an example of two heterogeneous social networks (Twitter and Foursquare) [9]. In each network, users are connected with each other through social links, and they are also connected with a set of locations, timestamps, and text contents through online activities. Moreover, some users have two accounts in two social networks separately, and they serve as anchor nodes to connect two networks. More generally, some interaction systems are too complex to be modeled as an HIN with a simple network schema. Knowledge graph [25] is such an example. We know that knowledge graph is based on resource description framework (RDF) data [21],

---

[2] www.imdb.com/.

[3] http://www.uspto.gov/patents/.

[4] http://www.douban.com/.

which complies with an $< Subject, Property, Object >$ model. Here, "Subject" and "Object" can be considered as objects, and "Property" can be considered as the relation between "Subject" and "Object". And thus a knowledge graph can be considered as a heterogeneous network, and such an example is shown in Fig. 1.4b. In such a semantic knowledge base, like Yago [26], there are more than 10-million entities (or nodes) of different types, and more than 120-million links among these entities. In such a schema-rich network, it is impossible to depict such network with a simple network schema.

In HIN, we distinguish the types of nodes and links, which should introduce some novel pattern discovery, compared to traditional homogeneous networks. Although many networked data can be modeled as heterogeneous networks, heterogeneous networks still have some limitations. Firstly, some real data are too complex to be modeled as meaningful HINs. For example, we can consider the RDF data as an HIN, while we cannot simply depict its network schema. Secondly, it may be difficult to analyze some networked data with an HIN perspective, even these data can be modeled as an HIN. These limitations are also the future works of HIN. We need to design more powerful mining methods in HIN to make it capable to be applied in more applications and discover more novel patterns.

1.4   Why Heterogeneous Information Network Analysis

In the past decades, link analysis has been extensively explored [4]. So many methods have been developed for information network analysis, and numerous data mining tasks have been explored in homogeneous networks, such as ranking, clustering, link prediction, and influence analysis. However, due to some unique characteristics (e.g., fusion of more information and rich semantics) of HIN, most methods in homogeneous networks cannot be directly applied in heterogeneous networks, and it is potential to discover more interesting patterns in this kind of networks.

It is a new development of data mining. Early data mining problems focused on analyzing feature vectors of objects. In the late 1990s, with the advent of WWW, more and more data mining researchers turned to studying links among objects. It is one of the main research directions to mine hidden patterns from feature and link information of objects. In these researches, homogeneous networks are usually constructed from interconnected objects. In recent years, abundant social media emerge, and many different types of objects are interconnected. It is hard to model these interacted objects as homogeneous networks, while it is natural to model different types of objects and relations among them as heterogeneous networks. Particularly, with the rapid increment of user-generated content online, big data analysis is an emergent yet important task to be studied. Variety is one significant characteristic of big data [35]. As a semi-structured representation, heterogeneous information network can be an effective way to model complex objects and their relations in big data.

It is an effective tool to fuse more information. Compared to homogeneous network, heterogeneous network is natural to fuse more objects and their interactions.

In addition, traditional homogeneous networks are usually constructed from single data source, while heterogeneous network can fuse information across multiple data sources. For example, customers use many services provided by Google, such as Google search, G-mail, maps, and Google+. So we can fuse this information with a heterogeneous information network, in which customers interact with many different types of objects, such as key words, mails, locations, and followers. Broadly speaking, heterogeneous information network can also fuse information cross multiple social network platforms [9]. We know that there are many social network platforms with different objectives, such as Facebook, Twitter, WeChat, and Weibo. Moreover, users often participate in multiple social networks. Since each social network only captures a partial or biased view of a user, we can fuse information across multiple social network platforms with multiple heterogeneous information networks, where each heterogeneous network represents information from one social network with some anchor nodes connecting these networks [38].

It contains rich semantics. In heterogeneous networks, different-typed objects and links coexist and they carry different semantic meanings. As a bibliographic example shown in Fig. 1.1, it includes author, paper, and venue object types. The relation type "Author-Paper" means authors writing papers, while the relation type "Paper-Venue" means papers published in venues. Considering the semantic information will lead to more subtle knowledge discovery. For example, in DBLP bibliographic data [29], if you find the most similar authors to "Christos Faloutsos," you will get his students, like Spiros Papadimitriou and Jimeng Sun, under the APA path, while the results are reputable researchers, like Jiawei Han and Rakesh Agrawal, under the $APVPA$ path. How to mine interesting patterns with the semantic information is a unique issue in heterogeneous networks.

References

1. Chakrabarti, S., et al.: Mining the Web: Analysis of Hypertext and Semi Structured Data. Morgan Kaufmann, San Francisco (2002)
2. Cook, D.J., Holder, L.B.: Graph-based data mining. IEEE Intell. Syst. **15**(2), 32–41 (2000)
3. Feldman, R.: Link analysis: current state of the art. In: Tutorial at the KDD-2 (2002)
4. Getoor, L., Diehl, C.P.: Link mining: a survey. SIGKDD Explor. **7**(2), 3–12 (2005)
5. Jamali, M., Lakshmanan, L.: HeteroMF: recommendation in heterogeneous information networks using context dependent factor models. In: WWW, pp. 643–654 (2013)
6. Jensen, D., Goldberg, H.: AAAI Fall Symposium on AI and Link Analysis. AAAI Press (1998)
7. Kim, J., Wilhelm, T.: What is a complex graph? Phys. A Stat. Mech. Appl. **387**(11), 2637–2652 (2008)
8. Kong, X., Cao, B., Yu, P.S.: Multi-label classification by mining label and instance correlations from heterogeneous information networks. In: KDD, pp. 614–622 (2013)
9. Kong, X., Zhang, J., Yu, P.S.: Inferring anchor links across multiple heterogeneous social networks. In: CIKM, pp. 179–188 (2013)
10. Kong, X., Yu, P.S., Ding, Y., Wild, D.J.: Meta path-based collective classification in heterogeneous information networks. In: CIKM, pp. 1567–1571 (2012)
11. Konstas, I., Stathopoulo, V., Jose, J.M.: On social networks and collaborative recommendation. In: SIGIR, pp. 195–202 (2009)

12. Lewis, T.G.: Network Science: Theory and Applications. Wiley, New York (2011)
13. Li, Y., Shi, C., Yu, P.S., Chen, Q.: HRank: a path based ranking method in heterogeneous information network. In: WAIM, pp. 553–565 (2014)
14. Liben-Nowell, D., Kleinberg, J.: The link prediction problem for social networks. In: CIKM, pp. 556–559 (2003)
15. Liu, J., Wang, C., Gao, J., Han, J.: Multi-view clustering via joint nonnegative matrix factorization. In: SDM, pp. 252–260 (2013)
16. Long, B., Zhang, Z.M., Yu, P.S.: Co-clustering by block value decomposition. In: KDD, pp. 635–640 (2005)
17. Long, B., Wu, X., Zhang, Z., Yu, P.S.: Unsupervised learning on k-partite graphs. In: KDD, pp. 317–326 (2006)
18. Long, B., Zhang, Z., Wu, X., Yu, P.S.: Spectral clustering for multi-type relational data. In: ICML, pp. 585–592 (2006)
19. Newman, M.E.: The structure and function of complex networks. SIAM Rev. **45**(2), 167–256 (2003)
20. Otte, E., Rousseau, R.: Social network analysis: a powerful strategy, also for the information sciences. J. Inf. Sci. **28**(6), 441–453 (2002)
21. Özsu, M.T.: A survey of RDF data management systems. Front. Comput. Sci. **10**(3), 418–432 (2016)
22. Shi, C., Kong, X., Yu, P.S., Xie, S., Wu, B.: Relevance search in heterogeneous networks. In: International Conference on Extending Database Technology, pp. 180–191 (2012)
23. Shi, C., Zhou, C., Kong, X., Yu, P.S., Liu, G., Wang, B.: HeteRecom: a semantic-based recommendation system in heterogeneous networks. In: KDD, pp. 1552–1555 (2012)
24. Shi, C., Zhang, Z., Luo, P., Yu, P.S., Yue, Y., Wu, B.: Semantic path based personalized recommendation on weighted heterogeneous information networks. In: The ACM International, pp. 453–462 (2015)
25. Singhal, A.: Introducing the Knowledge Graph: things, not strings. In: Official Google Blog (2012)
26. Suchanek, F.M., Kasneci, G., Weikum, G.: Yago: A core of semantic knowledge. In: WWW, pp. 697–706 (2007)
27. Sun, Y., Han, J.: Mining heterogeneous information networks: a structural analysis approach. SIGKDD Explor. **14**(2), 20–28 (2012)
28. Sun, Y., Yu, Y., Han, J.: Ranking-based clustering of heterogeneous information networks with star network schema. In: KDD, pp. 797–806 (2009)
29. Sun, Y., Han, J., Yan, X., Yu, P., Wu, T.: PathSim: meta path-based top-k similarity search in heterogeneous information networks. In: VLDB, pp. 992–1003 (2011)
30. Sun, Y., Norick, B., Han, J., Yan, X., Yu, P.S., Yu, X.: Integrating meta-path selection with user-guided object clustering in heterogeneous information networks. In: KDD, pp. 1348–1356 (2012)
31. Tang, L., Liu, H., Zhang, J., Nazeri, Z.: Community evolution in dynamic multi-mode networks. In: KDD, pp. 677–685 (2008)
32. Tang, J., Gao, H., Hu, X., Liu, H.: Exploiting homophily effect for trust prediction. In: WSDM, pp. 53–62 (2013)
33. Wang, R., Shi, C., Yu, P.S., Wu, B.: Integrating clustering and ranking on hybrid heterogeneous information network. In: PAKDD, pp. 583–594 (2013)
34. Wasserman, S.: Social Network Analysis: Methods and Applications. Cambridge University Press, Cambridge (1994)
35. Wu, X., Zhu, X., Wu, G., Ding, W.: Data mining with big data. IEEE Trans. Knowl. Data Eng. **26**(1), 97–107 (2014)
36. Yang, Y., Chawla, N.V., Sun, Y., Han, J.: Predicting links in multi-relational and heterogeneous networks. In: ICDM, pp. 755–764 (2012)
37. Yu, X., Ren, X., Sun, Y., Sturt, B., Khandelwal, U., Gu, Q., Norick, B., Han, J.: Recommendation in heterogeneous information networks with implicit user feedback. In: RecSys, pp. 347–350 (2013)

38. Zhang, J., Yu, P.S.: Integrated anchor and social link predictions across social networks. In: Proceedings of the 24th International Conference on Artificial Intelligence, IJCAI'15, pp. 2125–2131. AAAI Press (2015)
39. Zhong, E., Fan, W., Wang, J., Xiao, L., Li, Y.: ComSoc: adaptive transfer of user behaviors over composite social network. In: KDD, pp. 696–704 (2012)
40. Zhong, E., Fan, W., Zhu, Y., Yang, Q.: Modeling the dynamics of composite social networks. In: KDD, pp. 937–945 (2013)
41. Zhuang, H., Zhang, J., Brova, G., Tang, J., Cam, H., Yan, X., Han, J.: Mining query-based subnetwork outliers in heterogeneous information networks. In: ICDM, pp. 1127–1132 (2014)
42. Zou, L., Özsu, M.T., Chen, L., Shen, X., Huang, R., Zhao, D.: gStore: a graph-based SPARQL query engine. VLDB J. 23(4), 565–590 (2014)

Chapter 2
Survey of Current Developments

Abstract Heterogeneous information network (HIN) provides a new paradigm to manage networked data. Meanwhile, it also introduces new challenges for many data mining tasks. Here, we give a brief survey on recent developments of this field. Concretely, we have analyzed more than 100 referred papers published in the referred conferences and journals in recent years and divided them into seven categories according to their data mining tasks. In this chapter, we will summarize the developments on these seven main data mining tasks. Moreover, these data mining tasks are coarsely ordered from basic task to advanced task.

2.1 Similarity Search

Similarity measure is to evaluate the similarity of objects. It is the basis of many data mining tasks, such as Web search, clustering, and product recommendation. Similarity measure has been well studied for a long time. These studies can be roughly categorized into two types: feature-based approaches and link-based approaches. The feature-based approaches measure the similarity of objects based on their feature values, such as cosine similarity, Jaccard coefficient, and Euclidean distance. The link-based approaches measure the similarity of objects based on their link structures in a graph, such as Personalized PageRank [33] and SimRank [32].

Recently, many researchers begin to consider similarity measure on heterogeneous information networks. Different from similarity measure on homogeneous networks, similarity measure on HIN not only considers structure similarity of two objects but also takes the meta path connecting these two objects into account. As we know, there are different meta paths connecting two objects, and these meta paths contain different semantic meanings, which may lead to different similarities. And thus the similarity measure on HIN is meta path constraint.

Considering the semantics in meta paths constituted by different-typed objects, Sun et al. [88] first propose the path-based similarity measure PathSim to evaluate the similarity of same-typed objects based on symmetric paths. Following their work, some researchers [23, 24] extend PathSim by incorporating richer information, such as transitive similarity, temporal dynamics, and supportive attributes. A path-based

© Springer International Publishing AG 2017
C. Shi and P.S. Yu, *Heterogeneous Information Network Analysis and Applications*, Data Analytics, DOI 10.1007/978-3-319-56212-4_2

similarity join method [108] is proposed to return the top k similar pairs of objects based on user-specified join paths. Wang et al. [101] define a meta-path-based relation similarity measure, RelSim, to measure the similarity between relation instances in schema-rich HINs. In addition, Wang et al. [99] model a document as a heterogeneous information network and propose a novel similarity measure called KnowSim to compute the relevance of two documents. In information retrieval community, Lao and Cohen [46, 47] propose a path-constrained random walk (PCRW) model to measure the entity proximity in a labeled directed graph constructed by the rich metadata of scientific literature.

In order to evaluate the relevance of different-typed objects, Shi et al. [72, 74] propose HeteSim to measure the relevance of any object pair under arbitrary meta path. As an adaption of HeteSim, LSH-HeteSim [48] is proposed to mine the drug–target interaction in heterogeneous biological networks where drugs and targets are connected with complicated semantic paths. In order to overcome the shortcomings of HeteSim in high computation and memory demand, Meng et al. [62] propose the AvgSim measure that evaluates the similarity scores through two random walk processes along the given meta path and the reverse meta path, respectively. In addition, some methods [8, 141] combine meta path-based relevance search with user preference.

Although similarity measure based on meta path has shown the effectiveness in capturing the single relationship between source and target objects, such as the co-authorship under the meta path APA, it still has some shortcomings in some applications. For example, in bibliographic data, we would like to measure the relation of two authors based on the fact that their papers not only are published in the same conference but also have the same topics (i.e., the $APVPA$ and $APTPA$ paths). In order to measure the complex relevance between objects, Huang et al. [28] propose the relevance measure based on metastructure, which is a directed acyclic graph and can be considered as a combination of meta paths. Similarly, Fang et al. [20] identify metagraphs as a novel means to characterize the common structures for a desired class of proximity. Moreover, they propose a family of metagraph-based proximity and employ a supervised technique to automatically learn the right form of proximity within its family to suit the desired class.

More works begin to integrate the network structure and other information to measure similarity of objects in HIN. Combining the influence and similarity information, Wang et al. [102] simultaneously measure social influence and object similarity in a heterogeneous network to produce more meaningful similarity scores. Wang et al. [96] propose a model to learn relevance through analyzing the context of heterogeneous networks for online targeting. Yu et al. [116] predict the semantic meaning based on a user's query in the meta-path-based feature space and learn a ranking model to answer the similarity query. Recently, Zhang et al. [133] propose a similarity measure to compute similarity between centers in an x-star network according to the attribute similarities and the connections among centers.

2.2 Clustering

Clustering analysis is the process of partitioning a set of data objects (or observations) into a set of clusters, such that objects in a cluster are similar to one another, yet dissimilar to objects in other clusters. Conventional clustering is based on the features of objects, such as k-means and so on [30]. Recently, clustering based on networked data (e.g., community detection) has been studied a lot. This kind of methods models the data as a homogeneous network and uses the given measure (e.g., normalized cuts [78], and modularity [63]) to divide the network into a series of subgraphs. Many algorithms have been proposed to solve this NP-hard problem, such as spectral method, greedy method [93], and sampling technique [71]. Some researches also simultaneously consider objects' link structure and attribute information to increase the accuracy of clustering [110, 140].

Recently, clustering of heterogeneous networks attracts much attention. Compared with homogeneous networks, heterogeneous networks integrate multityped objects, which generates new challenges for clustering tasks. On the one hand, multiple types of objects coexisting in a network lead to new clustering paradigms. As a consequence, a cluster in HIN may include different types of objects sharing the same topic [82]. For example, in a bibliographic heterogeneous network, a cluster of the database area consists of a set of database authors, conferences, terms, and papers. In this way, clustering in HIN preserves richer information, but it also faces more challenges. On the other hand, abundant information contained in HIN makes it more convenient to integrate additional information or other learning tasks for clustering. In this section, we will review these works according to the types of integrated information or tasks.

The attribute information is widely integrated into clustering analysis on HIN. Aggarwal et al. [1] use the local succinctness property to create balanced communities across a heterogeneous network. Considering the incompleteness of objects' attributes and different types of links in heterogeneous information networks, Sun et al. [85] propose a model-based clustering algorithm to integrate the incomplete attribute information and the network structure information. Qi et al. [67] propose a clustering algorithm based on heterogeneous random fields to model the structure and content of social media networks with outlier links. Cruz et al. [14] integrate structural dimension and compositional dimension which composes an attributed graph to solve the community detection problem. Recently, a density-based clustering model TCSC [7] is proposed to detect clusters considering the connections in the network and the vertex attributes.

Text information plays an important role in many heterogeneous network studies. Deng et al. [17] introduce a topic model with biased propagation to incorporate heterogeneous information network with topic modeling in a unified way. Furthermore, they [16] propose a joint probabilistic topic model for simultaneously modeling the contents of multityped objects of a heterogeneous information network. LSA-PTM [103] is introduced to identify clusters of multityped objects by propagating the topics obtained by LSA on the HIN via the links between different objects. Incorporating both the document content and various links in the text related heterogeneous network, Wang et al. [104] propose a unified topic model for topic mining and multiple objects clustering. Recently, CHINC [98] uses general-purpose knowledge as indirect supervision to improve the clustering results.

User guide information is also integrated into the clustering analysis. Sun et al. [87] present a semi-supervised clustering algorithm to generate different clustering results with path selection according to user guidance. Luo et al. [58] firstly introduce the concept of relation-path to measure the similarity between same-typed objects and use the labeled information to weight relation-paths and then propose SemiRPClus for semi-supervised learning in HIN.

Clustering is usually an independent data mining task. However, it can be integrated with other mining tasks to improve performances through mutual enhancing. Recently, ranking-based clustering on heterogeneous information network has emerged, which shows its advantages on the mutual promotion of clustering and ranking. RankClus [83] generates clusters for a specified type of objects in a bitype network based on the idea that the qualities of clustering and ranking are mutually enhanced. The following work NetClus [82] is proposed to handle a network with the star-schema. Wang et al. [105] introduce ComClus to promote clustering and ranking performance by applying star-schema network with self-loop to combine the heterogeneous and homogeneous information. In addition, a general method HeProjI is proposed to do ranking-based clustering in heterogeneous networks with arbitrary schema by projecting the network into a sequence of subnetworks [75]. And Chen et al. [12] propose a probabilistic generative model to achieve clustering and ranking simultaneously on a heterogeneous network with arbitrary schema. To make use of both textual information and heterogeneous linked entities, Wang et al. [97] develop a clustering and ranking algorithm to construct multityped topical hierarchies automatically. What's more, Qiu et al. [68] propose an algorithm OcdRank to combine overlapping community detection and community-member ranking together in directed heterogeneous social networks.

Outlier detection is the process of finding data objects with behaviors that are very different from expectation. Outlier detection and clustering analysis are two highly related, but different-aimed tasks. To detect outliers, Gupta et al. [21] propose an outlier-aware approach based on joint nonnegative matrix factorization to discover popular community distribution patterns. Furthermore, they propose to detect association-based clique outliers in heterogeneous networks given a conjunctive select query [22]. What's more, Zhuang et al. [142] propose an outlier detection algorithm to find subnetwork outliers according to different queries and semantics. Also based on queries, Kuck et al. [44] propose a meta-path-based outlierness measure for mining outliers in heterogeneous networks.

In addition, some other information is also integrated. For example, a social influence-based clustering framework SI-Cluster is proposed to analyze heterogeneous information networks based on both people's connections and their social activities [138]. Besides the traditional models employed in clustering on HIN, such as topic model and spectral clustering, Alqadah et al. [4] propose a novel game theoretic framework for defining and mining clusters in heterogeneous information networks.

2.3 Classification

Classification is a data analysis task where a model or classifier is constructed to predict class (categorical) labels. Traditional machine learning has focused on the classification of identically structured objects satisfying independent identically distribution (IID). However, links exist among objects in many real-world datasets, which makes objects not satisfy IID. So link-based object classification has received considerable attention, where a data graph is composed of a set of objects connected to each other via a set of links. Many methods extend traditional classification methods to consider correlations among objects [45]. The link-based object classification usually considers that objects and links in the graph are identical, respectively. That is, the objects and links among them constitute a homogeneous network.

Different from traditional classification researches, the classification problems studied in HIN have some new characteristics. First, the objects contained in HIN are different-typed, which means we can classify multiple types of objects simultaneously. Second, label knowledge can spread through various links among different-typed objects. In the HIN condition, the label of objects is decided by the effects of different-typed objects along different-typed links.

Many works extend traditional classification to heterogeneous information networks. Some works extend transductive classification task, which is to predict labels for the given unlabeled data. For example, GNetMine [35] is proposed to model the link structure in information networks with arbitrary network schema and arbitrary number of object/link types. Wan et al. [94] propose a graph-regularized meta-path-based transductive regression model, which combines the principal philosophies of typical graph-based transductive classification methods and transductive regression models designed for homogeneous networks. Luo et al. propose HetPathMine [56] to cluster with small labeled data on HIN through a novel meta path selection model, and Jacob et al. [29] propose a method to label nodes of different types by computing a latent representation of nodes in a space where two connected nodes tend to have close latent representations. Recently, Bangcharoensap et al. [6] employ the edge betweenness centrality for the edge weight normalization and further improve the centrality to make it suitable for heterogeneous networks. Some works also extend inductive classification that is to construct a decision function in the whole data space. For example, Rossi et al. [70] use a bipartite heterogeneous network to represent textual document collections and propose IMBHN algorithm to induce a classification model assigning weights to textual terms.

Multilabel classification is prevalent in many real-world applications, where each example can be associated with a set of multiple labels simultaneously [41]. This kind of classification tasks is also extended to HIN. Angelova et al. [5] introduce a multilabel graph-based classification model for labeling heterogeneous networks by modeling the mutual influence between nodes as a random walk process. Kong et al. [41] use multiple types of relationships mined from the linkage structure of HIN to facilitate the multilabel classification process. Zhou et al. [139] propose an edge-centric multilabel classification approach considering both the structure affinity and the label vicinity.

As a unique characteristic, meta path is widely used in classification on HIN. Meta paths are usually used for feature generation in many methods, such as GNetMine [35] and HetPathMine [56]. Moreover, Kong et al. [40] introduce the concept of meta-path-based dependencies among objects to study the collective classification problem. Recently, Wang et al. [100] develop kernel methods based on meta paths in the HIN representation of texts for text classification.

Similar to clustering problem, classification is also integrated with other data mining tasks on HIN. Ranking-based classification is to integrate classification and ranking in a simultaneous, mutually enhancing process. Ji et al. [36] propose a ranking-based classification framework, RankClass, to perform more accurate analysis. As an extension of RankClass, Chen et al. [13] propose the F-RankClass for a unified classification framework that can be applied to binary or multiclass classification of unimodal or multimodal data. Some methods also integrate classification with information propagation. For example, Jendoubi et al. [34] classify the social message based on its spreading in the network and the theory of belief functions.

2.4 Ranking

Ranking is an important data mining task in network analysis, which evaluates object importance or popularity based on some ranking functions. Many ranking methods have been proposed in homogeneous networks, such as PageRank [65] and HITS [39]. These approaches only consider the same type of objects in homogeneous networks.

Ranking in heterogeneous information networks is an important and meaningful task, but faces several challenges. First, there are different types of objects and relations in HIN, and treating all objects equally will mix different types of objects together. Second, different types of objects and relations in HIN carry different semantic meanings, which may lead to different ranking results. Taking the bibliographic heterogeneous network as an example, ranking on authors may have different results under different meta paths [50, 77], since these meta paths will construct different link structures among authors. Moreover, the rankings of different-typed objects have mutual effects. For example, reputable authors usually publish papers on top conferences.

The co-ranking problem on bipartite graphs has been widely explored in the past decades. For example, Zhou et al. [137] co-rank authors and their publications by coupling two random walk processes, and co-HITS [15] incorporates the bipartite graph with the content information and the constraints of relevance. Soulier et al. [80] propose a bitype entity ranking algorithm to rank jointly documents and authors in a bibliographic network regarding a topical query by combining content-based and network-based features. There are also some ranking works on the multirelational network. For example, MultiRank [64] is proposed to determine the importance of both objects and relations simultaneously for multirelational data, and HAR [49] is proposed to determine hub and authority scores of objects and relevance scores of relations in multirelational data for query search. These two methods focus on the same type of objects with multirelations. Recently, Huang et al. [26] integrate both formal genre and inferred social networks with tweet networks to rank tweets. Although this work makes use of various types of objects in heterogeneous networks, it still ranks one type of objects.

Considering the characteristics of meta path on HIN, some works propose path-based ranking methods. For example, Liu et al. [54] develop a publication ranking method with pseudorelevance feedback by leveraging a number of meta paths on the heterogeneous bibliographic graph. Applying the tensor analysis, Li et al. [50] propose HRank to evaluate the importance of multiple types of objects and meta paths simultaneously.

Ranking problem is also extended to HIN constructed by social media network. For image search in social media, Tsai et al. [91] propose SocialRank which uses social hints for image search and ranking in social networks. To identify high-quality objects (questions, answers, and users) in Q&A systems, Zhang et al. [129] devise an unsupervised heterogeneous network-based framework to co-rank multiple objects in Q&A sites. For heterogeneous cross-domain ranking problem, Wang et al. [95] propose a general regularized framework to discover a latent space for two domains and minimize two weighted ranking functions simultaneously in the latent space. Considering the dynamic nature of literature networks, a mutual reinforcement ranking framework is proposed to rank the future popularity of new publications and young researchers simultaneously [106].

2.5 Link Prediction

Link prediction is a fundamental problem in link mining that attempts to estimate the likelihood of the existence of a link between two nodes, based on observed links and the attributes of nodes. Link prediction is often viewed as a simple binary classification problem: For any two potentially linked objects, predict whether the link exists (1) or not (0). One kind of approach is to make this prediction entirely based on structural properties of the network [51], and another kind of approach is to make use of attribute information for link prediction [66].

Link prediction in an HIN has been an important research topic for recent years, which has the following characteristics. First, the links to be predicted are of different types, since objects in HIN are connected with different types of links. Second, there are dependencies existing among multiple types of links. So link prediction in an HIN needs to predict multiple types of links collectively by capturing the diverse and complex relationships among different types of links and leveraging the complementary prediction information.

Utilizing the meta path, many works employ a two-step process to solve the link prediction problem in HIN. The first step is to extract meta path-based feature vectors, while the second step is to train a regression or classification model to compute the existence probability of a link [10, 11, 84, 86, 115]. For example, Sun et al. [84] propose PathPredict to solve the problem of co-author relationship prediction through meta path-based feature extraction and logistic regression-based model. Zhang et al. [130] use meta path-based features to predict organization chart or management hierarchy. Utilizing diverse and complex linkage information, Cao et al. [10] design a relatedness measure to construct the feature vectors of links and propose an iterative framework to predict multiple types of links collectively. In addition, Sun et al. [86] model the distribution of relationship building time with the use of the extracted topological features to predict when a certain relationship will be formed.

Probabilistic models are also widely applied for link prediction tasks in HIN. Yang et al. [112] propose a probabilistic method MRIP which models the influence propagating between heterogeneous relationships to predict links in multirelational heterogeneous networks. Also, the TFGM model [113] defines a latent topic layer to bridge multiple networks and designs a semi-supervised learning model to mine competitive relationships across heterogeneous networks. Dong et al. [19] develop a transfer-based ranking factor graph model that combines several social patterns with network structure information for link prediction and recommendation. Matrix factorization is another common tool to handle link prediction problems. For example, Huang et al. [27] develop the joint manifold factorization (JMF) method to perform trust prediction with the ancillary rating matrix via aggregating heterogeneous social networks.

The approaches mentioned above mainly focus on link prediction on one single heterogeneous network. Recently, Zhang et al. [42, 126, 128] propose the problem of link prediction across multiple aligned heterogeneous networks. A two-phase link prediction method is put forward in [42]. The first phase is to extract heterogeneous features from multiple networks, while the second phase is to infer anchor links by formulating it as a stable matching problem. In addition, Zhang et al. [126] propose SCAN-PS to solve the social link prediction problem for new users using the "anchors." Furthermore, they propose the TRAIL [128] method to predict social links and location links simultaneously. Also aimed at the cold-start problem of new users, Liu et al. [52] propose the aligned factor graph model for user–user link prediction problem by utilizing information from another similar social network. In order to identify users from multiple heterogeneous social networks and integrate different networks, an energy-based model COSNET [134] is proposed by considering both local and global consistency among multiple networks.

Most of the available works on link prediction are designed for static networks; however, the problem of dynamic link prediction is also very important and challenging. Taking into account both the dynamic and heterogeneous nature of Web data, Zhao et al. [135] propose a general framework to characterize and predict community members from the evolution of heterogeneous Web data. In order to solve the problem of dynamic link inference in temporal and heterogeneous information networks, Aggarwal et al. [2, 3] develop a two-level scheme which makes efficient macro- and microdecisions for combining the topology and type information. Aiming at predicting the distribution of the labels on neighbors of a given node, Ma et al. [60] propose an evolution factor model which utilizes two new structures, neighbor distribution vector and neighbor label evolution matrix.

2.6 Recommendation

Recommender system helps consumers to search products that are likely to be of interest to the user such as books, movies, and restaurants. It uses a broad range of techniques from information retrieval, statistics, and machine learning to search for similarities among items and customer preferences. Traditional recommended systems normally only utilize the user–item rating feedback information for recommendation. Collaborative filtering is one of the most popular techniques, which includes two types of approaches: memory-based methods and model-based methods. Recently, matrix factorization has shown its effectiveness and efficiency in recommended systems, which factorizes the user–item rating matrix into two low-rank user-specific and item-specific matrices and then utilizes the factorized matrices to make further predictions [81]. With the prevalence of social media, more and more researchers study social recommended system, which utilizes social relations among users [59, 111].

Recently, some researchers have begun to realize the importance of heterogeneous information for recommendations. The comprehensive information and rich semantics of HIN make it promising to generate better recommendations. For example, in an HIN extracted from movie-recommended system [76], it not only contains different types of objects (e.g., users and movies) but also illustrates all kinds of relations among objects, such as viewing information, social relations, and attribute information. Constructing heterogeneous networks for recommendation can effectively fuse all kinds of information, which can be potentially utilized for recommendation. Moreover, the objects and relations in the networks have different semantics, which can be explored to reveal subtle relations among objects.

Meta path is well used to explore the semantics and extract relations among objects. Shi et al. [73] implement a semantic-based recommended system, HeteRecom, which employs the semantics information of meta path to evaluate the similarities between movies. Furthermore, considering the attribute values, such as rating score on links, they model the recommended system as a weighted HIN and propose a semantic path-based personalized recommendation method SemRec [76]. In order to take full advantage of the relationship heterogeneity, Yu et al. [117, 118] introduce meta-path-based latent features to represent the connectivity between users and items along different types of paths and then define recommendation models at both global and personalized levels with Bayesian ranking optimization techniques. Also based on meta path, Burke et al. [9] present an approach for recommendation which incorporates multiple relations in a weighted hybrid.

A number of approaches employ heterogeneous information network to fuse various kinds of information. Utilizing different contexts information, Jamali et al. [31]

propose a context-dependent matrix factorization model which considers a general latent factor for every entity and context-dependent latent factors for every context. Using user implicit feedback data, Yu et al. [117, 118] solve the global and personalized entity recommendation problem. Based on related interest groups, Ren et al. [69] propose a cluster-based citation recommendation framework to predict each query's citations in bibliographic networks. Similarly, Wu et al. [107] exploit graph summarization and content-based clustering for media recommendation with the interest group information. Based on multiple heterogeneous network features, Yang et al. [109] model multiple features into a unified framework with a SVM-Rank-based method. And using multiple types of relations, Luo et al. [57] propose a social collaborative filtering algorithm. In addition, adopting the similarity of users and items as regularization, some works [75, 136] propose matrix factorization-based frameworks for recommendation.

2.7 Information Fusion

Information fusion denotes the process of merging information from heterogeneous sources with differing conceptual, contextual, and typographical representations. Due to the availability of various data sources, fusing these scattered distributed information sources has become an important research problem. In the past decades, dozens of papers have been published on this topic in many traditional data mining areas, e.g., data schema integration in data warehouse [61], protein–protein interaction (PPI) networks and gene regulatory networks matching in bioinformatics [79], and ontology mapping in Web semantics [18]. Nowadays, with the surge of HIN, information fusion across multiple HINs has become a novel yet important research problem. By fusing information from different HINs, we can obtain a more comprehensive and consistent knowledge about the common information entities shared in different HINs, including their structures, properties, and activities.

To fuse the information in multiple HINs, an important prerequisite will be to align the HINs via the shared common information entities, which can be users in social networks, authors in bibliographical networks, and protein molecules in biological networks. Perfect HIN alignment is a challenging problem as the underlying subgraph isomorphism problem is actually NP-complete [38]. Meanwhile, based on the structure and attribute information available in HINs, a large number of approximated HIN alignment algorithms have been proposed so far. Enlightened by the homogeneous network alignment method in [92], Koutra et al. [43] propose to align two bipartite graphs with a fast network alignment algorithm. Zafarani et al. [138] propose to match users across social networks based on various node attributes, e.g., username, typing patterns, and language patterns. Kong et al. [42] formulate the heterogeneous social network alignment problem as an anchor link prediction problem. A two-step supervised method MNA is proposed in [42] to infer potential anchor links across networks with heterogeneous information in the networks. However, social networks in the real world are actually mostly partially aligned, and lots of users are not anchor users. Zhang et al. have proposed the partial network alignment methods based on supervised learning setting and PU learning setting in [123, 131], respectively. In addition to these pairwise social network alignment problems, multiple (more than two) social networks can be aligned simultaneously. Zhang et al. [124] discover that the inferred cross-network mapping of entities in social network alignment should meet the transitivity law and has an inherent one-to-one constraint. A new multiple social alignment framework is introduced in [124] to minimize the alignment costs and preserve the transitivity law and one-to-one constraint on the inferred mappings. Besides users, many other kinds of information entities can also be shared by multiple social sites, such as the geo-spatial locations shared by location-based social networks and products shared by e-commerce sites. To infer the corresponding mapping between these different kinds of information entities simultaneously, Zhang et al. propose the network partial co-alignment problem in [125].

By fusing multiple HINs, the heterogeneous information available in each network can be transferred to other aligned networks, and lots of application problems on HIN, e.g., link prediction and friend recommendation [90, 123, 127], community detection [122], information diffusion [119, 120, 132], and product recommendation [55], will benefit from it a lot.

Via the inferred mappings, Zhang et al. propose to transfer heterogeneous links across aligned networks to improve quality of predicted links/recommended friends [123, 127]. Tang et al. [90] propose a transfer-based factor graph model which predicts the types of social relationships in a target network by borrowing knowledge from a different source network. For new networks [128] and new users [126] with little social activity information, the transferred information can greatly overcome the cold-start problem when predicting links for them. What's more, information about the shared entities across aligned networks can provide us with a more comprehensive knowledge about the community structures formed by them. By utilizing the information across multiple aligned networks, Zhang et al. [114] propose a new model to refine the clustering results of the shared entities with information in other aligned networks mutually. Jin et al. [37] propose a scalable framework to study the synergistic partitioning of multiple aligned large-scale networks, which takes the relationships among different networks into consideration and tries to maintain the consistency on partitioning the same nodes of different networks into the same partitions. Zhang et al. [122] study the community detection in emerging networks with information transferred from other aligned networks to overcome the cold-start problem. In addition, by fusing multiple heterogeneous social networks, users in networks will be extensively connected with each other via both intra-network connections (e.g., friendship connections among users) and inter-network connections (i.e., the inferred mappings across networks). As a result, information can reach more users and achieve broader influence across the aligned social networks. Zhan et al. propose a new model to study the information diffusion process across multiple aligned networks in [119] and introduce a new problem to discover the tipping users across aligned networks in [120].

2.8　Other Applications

Besides the tasks discussed above, there are many other applications in heterogeneous networks, such as influence propagation and privacy risk problem. To quantitatively learn influence from heterogeneous networks, Liu et al. [53] first use a generative graphical model to learn the direct influence and then use propagation methods to mine indirect and global influence. Using meta paths, Zhan et al. [119] propose a model M&M to solve the influence maximization problem in multiple partially aligned heterogeneous online social networks. For privacy risk in anonymized HIN, Zhang et al. [121] present a de-anonymization attack that exploits the identified vulnerability to prey upon the risk. Aiming at the inferior performances of unsupervised text embedding methods, Tang et al. [89] propose a semi-supervised representation learning method for text data, in which labeled information and different levels of word co-occurrence information are represented as a large-scale heterogeneous text network. To improve the effectiveness of offline sales, Hu et al. [25] construct a company-to-company graph from semantics based meta-path learning and then adopt label propagation on the graph to predict promising companies.

References

1. Aggarwal, C., Xie, Y., Yu, P.: Towards community detection in locally heterogeneous networks. In: SDM, pp. 391–402 (2011)
2. Aggarwal, C.C., Xie, Y., Yu, P.S.: On dynamic link inference in heterogeneous networks. In: SDM, pp. 415–426 (2012)
3. Aggarwal, C.C., Xie, Y., Yu, P.S.: A framework for dynamic link prediction in heterogeneous networks. Stat. Anal. Data Min. ASA Data Sci. J. **7**(1), 14–33 (2014)
4. Alqadah, F., Bhatnagar, R.: A game theoretic framework for heterogeneous information network clustering. In: KDD, pp. 795–802 (2011)
5. Angelova, R., Kasneci, G., Weikum, G.: Graffiti: graph-based classification in heterogeneous networks. In: WWW, pp. 139–170 (2012)
6. Bangcharoensap, P., Murata, T., Kobayashi, H., Shimizu, N.: Transductive classification on heterogeneous information networks with edge betweenness-based normalization. In: WSDM, pp. 437–446 (2016)
7. Boden, B., Ester, M., Seidl, T.: Density-based subspace clustering in heterogeneous networks. In: ECML/PKDD, pp. 149–164 (2014)
8. Bu, S., Hong, X., Peng, Z., Li, Q.: Integrating meta-path selection with user-preference for top-k relevant search in heterogeneous information networks. In: CSCWD, pp. 301–306 (2014)
9. Burke, R., Vahedian, F., Mobasher, B.: Hybrid recommendation in heterogeneous networks. In: UMAP, pp. 49–60 (2014)
10. Cao, B., Kong, X., Yu, P.S.: Collective prediction of multiple types of links in heterogeneous information networks. In: ICDM, pp. 50–59 (2014)
11. Chen, J., Gao, H., Wu, Z., Li, D.: Tag co-occurrence relationship prediction in heterogeneous information networks. In: ICPADS, pp. 528–533 (2013)
12. Chen, J., Dai, W., Sun, Y., Dy, J.: Clustering and ranking in heterogeneous information networks via gamma-poisson model. In: SDM, pp. 425–432 (2015)
13. Chen, S.D., Chen, Y.Y., Han, J., Moulin, P.: A feature-enhanced ranking-based classifier for multimodal data and heterogeneous information networks. In: ICDM, pp. 997–1002 (2013)

14. Cruz, J.D., Bothorel, C., Poulet, F.: Integrating heterogeneous information within a social network for detecting communities. In: ASONAM, pp. 1453–1454 (2013)
15. Deng, H., Lyu, M.R., King, I.: A generalized Co-HITS algorithm and its application to bipartite graphs. In: KDD, pp. 239–248 (2009)
16. Deng, H., Zhao, B., Han, J.: Collective topic modeling for heterogeneous networks. In: SIGIR, pp. 1109–1110 (2011)
17. Deng, H., Han, J., Zhao, B., Yu, Y., Lin, C.X.: Probabilistic topic models with biased propagation on heterogeneous information networks. In: KDD, pp. 795–802 (2011)
18. Doan, A., Madhavan, J., Domingos, P., Halevy, A.: Ontology matching: a machine learning approach. Handbook on Ontologies, pp. 385–403. Springer, Berlin (2004)
19. Dong, Y., Tang, J., Wu, S., Tian, J., Chawla, N.V., Rao, J., Cao, H.: Link prediction and recommendation across heterogeneous social networks. In: ICDM, pp. 181–190 (2012)
20. Fang, Y., Lin, W., Zheng, V.W., Wu, M., Chang, C.C., Li, X.L.: Semantic proximity search on graphs with metagraph-based learning. In: ICDE, pp. 277–288 (2016)
21. Gupta, M., Gao, J., Han, J.: Community distribution outlier detection in heterogeneous information networks. In: ECML, pp. 557–573 (2013)
22. Gupta, M., Gao, J., Yan, X., Cam, H., Han, J.: On detecting association-based clique outliers in heterogeneous information networks. In: ASONAM, pp. 108–115 (2013)
23. He, J., Bailey, J., Zhang, R.: Exploiting transitive similarity and temporal dynamics for similarity search in heterogeneous information networks. In: International Conference on Database Systems for Advanced Applications, pp. 141–155 (2014)
24. Hou U.L., Yao, K., Mak, H.: PathSimExt: revisiting PathSim in heterogeneous information networks. In: WAIM, pp. 38–42 (2014)
25. Hu, Q., Xie, S., Zhang, J., Zhu, Q., Guo, S., Yu, P.S.: HeteroSales: utilizing heterogeneous social networks to identify the next enterprise customer. In: WWW, pp. 41–50 (2016)
26. Huang, H., Zubiaga, A., Ji, H., Deng, H., Wang, D., Le, H.K., Abdelzaher, T.F., Han, J., Leung, A., Hancock, J.P., Others: Tweet ranking based on heterogeneous networks. In: COLING, pp. 1239–1256 (2012)
27. Huang, J., Nie, F., Huang, H., Tu, Y.C.: Trust prediction via aggregating heterogeneous social networks. In: CIKM, pp. 1774–1778 (2012)
28. Huang, Z., Zheng, Y., Cheng, R., Sun, Y., Mamoulis, N., Li, X.: Meta structure: computing relevance in large heterogeneous information networks. In: SIGKDD, pp. 1595–1604 (2016)
29. Jacob, Y., Denoyer, L., Gallinari, P.: Learning latent representations of nodes for classifying in heterogeneous social networks. In: WSDM, pp. 373–382 (2014)
30. Jain, A.K.: Data clustering: 50 years beyond K-means. Pattern Recognit. Lett. **31**(8), 651–666 (2010)
31. Jamali, M., Lakshmanan, L.: HeteroMF: recommendation in heterogeneous information networks using context dependent factor models. In: WWW, pp. 643–654 (2013)
32. Jeh, G., Widom, J.: SimRank: a measure of structural-context similarity. In: KDD, pp. 538–543 (2002)
33. Jeh, G., Widom, J.: Scaling personalized web search. In: WWW, pp. 271–279 (2003)
34. Jendoubi, S., Martin, A., Lietard, L., Yaghlane, B.B.: Classification of message spreading in a heterogeneous social network. In: IPMU, pp. 66–75 (2014)
35. Ji, M., Sun, Y., Danilevsky, M., Han, J., Gao, J.: Graph regularized transductive classification on heterogeneous information networks. In: ECML/PKDD, pp. 570–586 (2010)
36. Ji, M., Han, J., Danilevsky, M.: Ranking-based classification of heterogeneous information networks. In: KDD, pp. 1298–1306 (2011)
37. Jin, S., Zhang, J., Yu, P.S., Yang, S., Li, A.: Synergistic partitioning in multiple large scale social networks. In: IEEE BigData, pp. 281–290 (2014)
38. Klau, G.W.: A new graph-based method for pairwise global network alignment. BMC Bioinform. **10**(Suppl 1), S59 (2009)
39. Kleinberg, J.M.: Authoritative sources in a hyperlinked environment. In: SODA, pp. 668–677 (1999)

40. Kong, X., Yu, P.S., Ding, Y., Wild, D.J.: Meta path-based collective classification in heterogeneous information networks. In: CIKM, pp. 1567–1571 (2012)
41. Kong, X., Cao, B., Yu, P.S.: Multi-label classification by mining label and instance correlations from heterogeneous information networks. In: KDD, pp. 614–622 (2013)
42. Kong, X., Zhang, J., Yu, P.S.: Inferring anchor links across multiple heterogeneous social networks. In: CIKM, pp. 179–188 (2013)
43. Koutra, D., Tong, H., Lubensky, D.: Big-align: fast bipartite graph alignment. In: ICDM, pp. 389–398 (2013)
44. Kuck, J., Zhuang, H., Yan, X., Cam, H., Han, J.: Query-based outlier detection in heterogeneous information networks. In: EDBT, pp. 325–336 (2015)
45. Lafferty, J., McCallum, A., Pereira, F.C.N.: Conditional random fields: probabilistic models for segmenting and labeling sequence data. In: ICML, pp. 282–289 (2001)
46. Lao, N., Cohen, W.: Fast query execution for retrieval models based on path constrained random walks. In: KDD, pp. 881–888 (2010)
47. Lao, N., Cohen, W.W.: Relational retrieval using a combination of path-constrained random walks. Mach. Learn. **81**(2), 53–67 (2010)
48. Li, C., Sun, J., Xiong, Y., Zheng, G.: An efficient drug-target interaction mining algorithm in heterogeneous biological networks. In: PAKDD, pp. 65–76 (2014)
49. Li, X., Ng, M.K., Ye, Y.: HAR: hub, authority and relevance scores in multi-relational data for query search. In: SDM, pp. 141–152 (2012)
50. Li, Y., Shi, C., Yu, P.S., Chen, Q.: HRank: a path based ranking method in heterogeneous information network. In: WAIM, pp. 553–565 (2014)
51. Liben-Nowell, D., Kleinberg, J.: The link-prediction problem for social networks. J. Am. Soc. Inf. Sci. Tech. **58**(7), 1019–1031 (2007)
52. Liu, F., Xia, S.: Link prediction in aligned heterogeneous networks. In: PAKDD, pp. 33–44 (2015)
53. Liu, L., Tang, J., Han, J., Yang, S.: Learning influence from heterogeneous social networks. Data Min. Knowl. Discov. **25**(3), 511–544 (2012)
54. Liu, X., Yu, Y., Guo, C., Sun, Y.: Meta-path-based ranking with pseudo relevance feedback on heterogeneous graph for citation recommendation. In: CIKM, pp. 121–130 (2014)
55. Lu, C.T., Xie, S., Shao, W., He, L., Yu, P.S.: Item recommendation for emerging online businesses. In: IJCAI, pp. 3797–3803 (2016)
56. Luo, C., Guan, R., Wang, Z., Lin, C.: HetPathMine: a novel transductive classification algorithm on heterogeneous information networks. In: Advances in Information Retrieval, vol. 8416, pp. 210–221 (2014)
57. Luo, C., Pang, W., Wang, Z.: Hete-CF: social-based collaborative filtering recommendation using heterogeneous relations. In: ICDM, pp. 917–922 (2014)
58. Luo, C., Pang, W., Wang, Z.: Semi-supervised clustering on heterogeneous information networks. In: Advances in Knowledge Discovery and Data Mining, vol. 8444, pp. 548–559 (2014)
59. Ma, H., King, I., Lyu, M.R.: Learning to recommend with social trust ensemble. In: SIGIR, pp. 203–210 (2009)
60. Ma, Y., Yang, N., Li, C., Zhang, L., Yu, P.S.: Predicting neighbor distribution in heterogeneous information networks. In: SDM, pp. 784–791 (2015)
61. Melnik, S., Garcia-Molina, H., Rahm, E.: Similarity flooding: a versatile graph matching algorithm and its application to schema matching. In: ICDE, pp. 117–128 (2002)
62. Meng, X., Shi, C., Li, Y., Zhang, L., Wu, B.: Relevance measure in large-scale heterogeneous networks. In: APWeb, pp. 636–643 (2014)
63. Newman, M.E.J., Girvan, M., M.E.J., Newman, M.G.: Finding and evaluating community structure in networks. Phys. Rev. E **69**(026113), 1757–1771 (2004)
64. Ng, M.K., Li, X., Ye, Y., Ng, M., Li, X., Ye, Y.: MultiRank: co-ranking for objects and relations in multi-relational data. In: KDD, pp. 1217–1225 (2011)
65. Page, L., Brin, S., Motwani, R., Winograd, T.: The pagerank citation ranking: bringing order to the web. In: Stanford InfoLab, pp. 1–14 (1998)

66. Popescul, A., Ungar, L.H.: Statistical relational learning for link prediction. In: IJCAI Workshop on Learning Statistical Models from Relational Data, vol. 2003 (2003)
67. Qi, G.J., Aggarwal, C.C., Huang, T.S.: On clustering heterogeneous social media objects with outlier links. In: WSDM, pp. 553–562 (2012)
68. Qiu, C., Chen, W., Wang, T., Lei, K.: Overlapping community detection in directed heterogeneous social network. In: WAIM, pp. 490–493 (2015)
69. Ren, X., Liu, J., Yu, X., Khandelwal, U., Gu, Q., Wang, L., Han, J.: ClusCite: effective citation recommendation by information network-based clustering. In: KDD, pp. 821–830 (2014)
70. Rossi, R.G., de Paulo Faleiros, T., de Andrade Lopes, A., Rezende, S.O.: Inductive model generation for text categorization using a bipartite heterogeneous network. In: ICDM, pp. 1086–1091 (2012)
71. Sales-Pardo, M., Guimera, R., Moreira, A.A., Amaral, L.A.N.: Extracting the hierarchical organization of complex systems. Proc. Natl. Acad. Sci. **104**(39), 15224–15229 (2007)
72. Shi, C., Kong, X., Yu, P.S., Xie, S., Wu, B.: Relevance search in heterogeneous networks. In: EDBT, pp. 180–191 (2012)
73. Shi, C., Zhou, C., Kong, X., Yu, P.S., Liu, G., Wang, B.: HeteRecom: a semantic-based recommendation system in heterogeneous networks. In: KDD, pp. 1552–1555 (2012)
74. Shi, C., Kong, X., Huang, Y., Philip, S.Y., Wu, B.: Hetesim: a general framework for relevance measure in heterogeneous networks. IEEE Trans. Knowl. Data Eng. **26**(10), 2479–2492 (2014)
75. Shi, C., Wang, R., Li, Y., Yu, P.S., Wu, B.: Ranking-based clustering on general heterogeneous information networks by network projection. In: CIKM, pp. 699–708 (2014)
76. Shi, C., Zhang, Z., Luo, P., Yu, P.S., Yue, Y., Wu, B.: Semantic path based personalized recommendation on weighted heterogeneous information networks. In: CIKM, pp. 453–462 (2015)
77. Shi, C., Li, Y., Philip, S.Y., Wu, B.: Constrained-meta-path-based ranking in heterogeneous information network. Knowl. Inf. Syst. 1–29 (2016)
78. Shi, J., Malik, J.: Normalized cuts and image segmentation. IEEE Trans. Pattern Anal. Mach. Intell. **22**(8), 888–905 (2000)
79. Shih, Y.K., Parthasarathy, S.: Scalable global alignment for multiple biological networks. BMC Bioinf. **13**, 1–13 (2012)
80. Soulier, L., Jabeur, L.B., Tamine, L., Bahsoun, W.: On ranking relevant entities in heterogeneous networks using a language-based model. J. Am. Soc. Inf. Sci. Technol. **64**(3), 500–515 (2013)
81. Srebro, N., Jaakkola, T.: Weighted low-rank approximations. In: ICML, pp. 720–727 (2003)
82. Sun, Y., Yu, Y., Han, J.: Ranking-based clustering of heterogeneous information networks with star network schema. In: Proceedings of the 15th ACM SIGKDD International Conference on Knowledge Discovery and Data Mining, pp. 797–806 (2009)
83. Sun, Y., Han, J., Zhao, P., Yin, Z., Cheng, H., Wu, T.: RankClus: integrating clustering with ranking for heterogeneous information network analysis. In: EDBT, pp. 565–576 (2009)
84. Sun, Y., Barber, R., Gupta, M., Aggarwal, C.C., Han, J.: Co-author relationship prediction in heterogeneous bibliographic networks. In: ASONAM, pp. 121–128 (2011)
85. Sun, Y., Aggarwal, C., Han, J.: Relation strength-aware clustering of heterogeneous information networks with incomplete attributes. In: VLDB, pp. 394–405 (2012)
86. Sun, Y., Han, J., Aggarwal, C.C., Chawla, N.V.: When will it happen?: relationship prediction in heterogeneous information networks. In: WSDM, pp. 663–672 (2012)
87. Sun, Y., Norick, B., Han, J., Yan, X., Yu, P.S., Yu, X.: Integrating meta-path selection with user-guided object clustering in heterogeneous information networks. In: KDD, pp. 1348–1356 (2012)
88. Sun, Y.Z., Han, J.W., Yan, X.F., Yu, P.S., Wu, T.: PathSim: meta path-based Top-K similarity search in heterogeneous information networks. In: VLDB, pp. 992–1003 (2011)
89. Tang, J., Qu, M., Mei, Q.: PTE: predictive text embedding through large-scale heterogeneous text networks. In: KDD, pp. 1165–1174 (2015)
90. Tang, J., Lou, T., Kleinberg, J., Wu, S.: Transfer learning to infer social ties across heterogeneous networks. ACM Trans. Inf. Syst. **34**(2), 7:1–7:43 (2016)

91. Tsai, M.H., Aggarwal, C., Huang, T.: Ranking in heterogeneous social media. In: WSDM, pp. 613–622 (2014)
92. Umeyama, S.: An eigendecomposition approach to weighted graph matching problems. IEEE Trans. Pattern Anal. Mach. Intell. **10**(5), 695–703 (1988)
93. Wakita, K., Tsurumi, T.: Finding community structure in mega-scale social networks. In: WWW, pp. 1275–1276 (2007)
94. Wan, M., Ouyang, Y., Kaplan, L., Han, J.: Graph regularized meta-path based transductive regression in heterogeneous information network. In: SDM, pp. 918–926 (2015)
95. Wang, B., Tang, J., Fan, W., Chen, S., Tan, C., Yang, Z.: Query-dependent cross-domain ranking in heterogeneous network. Knowl. Inf. Syst. **34**(1), 109–145 (2013)
96. Wang, C., Raina, R., Fong, D., Zhou, D., Han, J., Badros, G.J.: Learning relevance from heterogeneous social network and its application in online targeting. In: SIGIR, pp. 655–664 (2011)
97. Wang, C., Danilevsky, M., Liu, J., Desai, N., Ji, H., Han, J.: Constructing topical hierarchies in heterogeneous information networks. In: ICDM, pp. 767–776 (2013)
98. Wang, C., Song, Y., El-Kishky, A., Roth, D., Zhang, M., Han, J.: Incorporating world knowledge to document clustering via heterogeneous information networks. In: KDD, pp. 1215–1224 (2015)
99. Wang, C., Song, Y., Li, H., Zhang, M., Han, J.: Knowsim: a document similarity measure on structured heterogeneous information networks. In: ICDM, pp. 1015–1020 (2015)
100. Wang, C., Song, Y., Li, H., Zhang, M., Han, J.: Text classification with heterogeneous information network kernels. In: AAAI, pp. 2130–2136 (2016)
101. Wang, C., Sun, Y., Song, Y., Han, J., Song, Y., Wang, L., Zhang, M.: Relsim: relation similarity search in schema-rich heterogeneous information networks. In: Siam International Conference on Data Mining, pp. 621–629 (2016)
102. Wang, G., Hu, Q., Yu, P.S.: Influence and similarity on heterogeneous networks. In: CIKM, pp. 1462–1466 (2012)
103. Wang, Q., Peng, Z., Jiang, F., Li, Q.: LSA-PTM: a propagation-based topic model using latent semantic analysis on heterogeneous information networks. In: WAIM, pp. 13–24 (2013)
104. Wang, Q., Peng, Z., Wang, S., Yu, P.S., Li, Q., Hong, X.: cluTM: content and link integrated topic model on heterogeneous information networks. In: WAIM, pp. 207–218 (2015)
105. Wang, R., Shi, C., Yu, P.S., Wu, B.: Integrating clustering and ranking on hybrid heterogeneous information network. In: PAKDD, pp. 583–594 (2013)
106. Wang, S., Xie, S., Zhang, X., Li, Z., Yu, P.S., Shu, X.: Future influence ranking of scientific literature. In: SDM, pp. 749–757 (2014)
107. Wu, J., Chen, L., Yu, Q., Han, P., Wu, Z.: Trust-aware media recommendation in heterogeneous social networks. WWW **18**(1), 139–157 (2015)
108. Xiong, Y., Zhu, Y., Yu, P.S.: Top-k similarity join in heterogeneous information networks. IEEE Trans. Knowl. Data Eng. **27**(6), 1710–1723 (2015)
109. Yang, C., Sun, J., Ma, J., Zhang, S., Wang, G., Hua, Z.: Scientific collaborator recommendation in heterogeneous bibliographic networks. In: HICSS, pp. 552–561 (2015)
110. Yang, T., Jin, R., Chi, Y., Zhu, S.: Combining link and content for community detection: a discriminative approach. In: KDD, pp. 927–936 (2009)
111. Yang, X., Steck, H., Liu, Y.: Circle-based recommendation in online social networks. In: KDD, pp. 1267–1275 (2012)
112. Yang, Y., Chawla, N.V., Sun, Y., Han, J.: Predicting links in multi-relational and heterogeneous networks. In: ICDM, pp. 755–764 (2012)
113. Yang, Y., Tang, J., Keomany, J., Zhao, Y., Li, J., Ding, Y., Li, T., Wang, L.: Mining competitive relationships by learning across heterogeneous networks. In: CIKM, pp. 1432–1441 (2012)
114. Yu, P.S., Zhang, J.: MCD: mutual clustering across multiple social networks. In: IEEE International Congress on Big Data, pp. 762–771 (2015)
115. Yu, X., Gu, Q., Zhou, M., Han, J.: Citation prediction in heterogeneous bibliographic networks. In: SDM, pp. 1119–1130 (2012)

116. Yu, X., Sun, Y., Norick, B., Mao, T., Han, J.: User guided entity similarity search using meta-path selection in heterogeneous information networks. In: CIKM, pp. 2025–2029 (2012)
117. Yu, X., Ren, X., Sun, Y., Sturt, B., Khandelwal, U., Gu, Q., Norick, B., Han, J.: Recommendation in heterogeneous information networks with implicit user feedback. In: RecSys, pp. 347–350 (2013)
118. Yu, X., Ren, X., Sun, Y., Gu, Q., Sturt, B., Khandelwal, U., Norick, B., Han, J.: Personalized entity recommendation: a heterogeneous information network approach. In: WSDM, pp. 283–292 (2014)
119. Zhan, Q., Zhang, J., Wang, S., Yu, P.S., Xie, J.: Influence maximization across partially aligned heterogeneous social networks. In: PAKDD, pp. 58–69 (2015)
120. Zhan, Q., Zhang, J., Philip, S.Y., Emery, S., Xie, J.: Discover tipping users for cross network influencing. In: 2016 IEEE 17th International Conference on Information Reuse and Integration (IRI), pp. 67–76 (2016)
121. Zhang, A., Xie, X., Chang, K.C.C., Gunter, C.A., Han, J., Wang, X.: Privacy risk in anonymized heterogeneous information networks. In: EDBT, pp. 595–606 (2014)
122. Zhang, J., Yu, P.: Community detection for emerging networks. In: SDM, pp. 127–135 (2015)
123. Zhang, J., Yu, P.S.: Integrated anchor and social link predictions across social networks. In: IJCAI, pp. 2125–2131 (2015)
124. Zhang, J., Yu, P.S.: Multiple anonymized social networks alignment. In: ICDM, pp. 599–608 (2015)
125. Zhang, J., Yu, P.S.: PCT: partial co-alignment of social networks. In: WWW, pp. 749–759 (2016)
126. Zhang, J., Kong, X., Yu, P.S.: Predicting social links for new users across aligned heterogeneous social networks. In: ICDM, pp. 1289–1294 (2013)
127. Zhang, J., Yu, P.S., Zhou, Z.H.: Meta-path based multi-network collective link prediction. In: KDD, pp. 1286–1295 (2014)
128. Zhang, J., Kong, X., Yu, P.S.: Transferring heterogeneous links across location-based social networks. In: WSDM, pp. 303–312 (2014)
129. Zhang, J., Kong, X., Jie, L., Chang, Y., Yu, P.S.: NCR: a scalable network-based approach to co-ranking in question-and-answer sites. In: CIKM, pp. 709–718 (2014)
130. Zhang, J., Yu, P.S., Lv, Y.: Organizational chart inference. In: KDD, pp. 1435–1444 (2015)
131. Zhang, J., Shao, W., Wang, S., Kong, X., Yu, P.S.: Partial network alignment with anchor meta path and truncated generic stable matching. ArXiv e-prints (2015)
132. Zhang, J., Yu, P.S., Lv, Y., Zhan, Q.: Information diffusion at workplace. In: CIKM, pp. 1673–1682. ACM (2016)
133. Zhang, M., Hu, H., He, Z., Wang, W.: Top-k similarity search in heterogeneous information networks with x-star network schema. Expert Syst. Appl. **42**(2), 699–712 (2015)
134. Zhang, Y., Tang, J., Yang, Z., Pei, J., Yu, P.S.: COSNET: connecting heterogeneous social networks with local and global consistency. In: KDD, pp. 1485–1494 (2015)
135. Zhao, Q., Bhowmick, S.S., Zheng, X., Yi, K.: Characterizing and predicting community members from evolutionary and heterogeneous networks. In: CIKM, pp. 309–318 (2008)
136. Zheng, J., Liu, J., Shi, C., Zhuang, F., Li, J., Wu, B.: Dual similarity regularization for recommendation. In: PAKDD, pp. 542–554 (2016)
137. Zhou, D., Orshanskiy, S.A., Zha, H., Giles, C.L.: Co-ranking authors and documents in a heterogeneous network. In: ICDM, pp. 739–744 (2007)
138. Zhou, Y., Liu, L.: Social influence based clustering of heterogeneous information networks. In: KDD, pp. 338–346 (2013)
139. Zhou, Y., Liu, L.: Activity-edge centric multi-label classification for mining heterogeneous information networks. In: KDD, pp. 1276–1285 (2014)

140. Zhou, Y., Cheng, H., Yu, J.X.: Graph clustering based on structural/attribute similarities. In: VLDB, pp. 718–729 (2009)
141. Zhu, M., Zhu, T., Peng, Z., Yang, G., Xu, Y., Wang, S., Wang, X., Hong, X.: Relevance search on signed heterogeneous information network based on meta-path factorization. In: WAIM, pp. 181–192 (2015)
142. Zhuang, H., Zhang, J., Brova, G., Tang, J., Cam, H., Yan, X., Han, J.: Mining query-Based subnetwork outliers in heterogeneous information networks. In: ICDM, pp. 1127–1132 (2014)

Chapter 3
Relevance Measure of Heterogeneous Objects

Abstract Similarity search is an important function in many applications, which usually focuses on measuring the similarity between objects with the same type. However, in many scenarios, we need to measure the relatedness between objects with different types. With the surge of study on heterogeneous networks, the relevance measure on objects with different types becomes increasingly important. In this chapter, we study the relevance search problem in heterogeneous networks, where the task is to measure the relatedness of heterogeneous objects (including objects with the same type or different types). And then, we introduce a novel measure HeteSim and its extended version.

3.1 HeteSim: A Uniform and Symmetric Relevance Measure

3.1.1 Overview

Similarity search is an important task in a wide range of applications, such as Web search [15] and product recommendations [11]. The key of similarity search is similarity measure, which evaluates the similarity of object pairs. Similarity measure has been extensively studied for traditional categorical and numerical data types, such as Jaccard coefficient and cosine similarity. There are also a few studies on leveraging link information in networks to measure the node similarity, such as Personalized PageRank [7], SimRank [6], and PathSim [21]. Conventional study on the similarity measure focuses on objects with the same type. That is, the objects being measured are of the same type, such as "document-to-document" and "Webpage-to-Webpage." There are very few studies on similarity measure on objects with different types. That is, the objects being measured are of different types, such as "author-to-conference" and "user-to-movie." It is reasonable. The similarity of objects with different types is a little against our common sense. Moreover, different from the similarity of objects with the same type, which can be measured on homogeneous situation (e.g., the same feature space or homogeneous link structure), it is even harder to define the similarity of objects with different types.

© Springer International Publishing AG 2017
C. Shi and P.S. Yu, *Heterogeneous Information Network Analysis and Applications*, Data Analytics, DOI 10.1007/978-3-319-56212-4_3

However, the similarity of objects with different types is not only meaningful but also useful in some scenarios. For example, Prof. Jiawei Han is more relevant to KDD than IJCAI. Moreover, the similarity measure of objects with different types is needed in many applications. For example, in a recommended system, we need to know the relatedness between users and items to make accurate recommendations [5]. In an automatic profile extraction application, we need to measure the relatedness of objects with different types, such as authors and conferences, and conferences and organizations. Particularly, with the advent of study on heterogeneous information networks [20, 21], it is not only increasingly important but also feasible to study the relatedness among objects with different types. Heterogeneous information networks are the logical networks involving multiple-typed objects and multiple-typed links denoting different relations [4]. It is clear that heterogeneous information networks are ubiquitous and form a critical component of modern information infrastructure [4]. So it is essential to provide a relevance search function on objects with different types in such networks, which is the base of many applications. Since objects with different types coexist in the same network, their relevance measure is possible through link structure.

In this chapter, we study the relevance search problem in heterogeneous information networks. The aim of relevance search is to effectively measure the relatedness of heterogeneous objects (including objects with the same type or different types). Different from the similarity search which measures only the similarity of objects with the same type, the relevance search measures the relatedness among heterogeneous objects and it is not limited to objects with the same type. Distinct from relational retrieval [13, 23] in information retrieval domain, here relevance search is done on heterogeneous networks which can be constructed from metadata of objects. Moreover, we think that a desirable relevance measure should satisfy the symmetry property based on the following reasons: (1) The symmetric measure is more general and useful in many learning tasks. Although the symmetry property is not necessary in the query task, it is essential for many important tasks, such as clustering and collaborative filtering. Moreover, it is the necessary condition for a metric. (2) The symmetric measure makes more sense in many applications, especially for the relatedness of heterogeneous object pairs. For example, in some applications, we need to answer the question like who has similar importance to the SIGIR conference as Jiawei Han to KDD. Through comparing the relatedness of object pairs, we can deduce the information of their relative importance. However, it can only be done by the symmetric measure, not the asymmetric measure.

Inspired by the intuition that two objects are related if they are referenced by related objects, we propose a general framework, called HeteSim, to evaluate the relatedness of heterogeneous objects in heterogeneous networks. HeteSim is a path-based relevance measure, which can effectively capture the subtle semantics of search paths. Based on pairwise random walk model, HeteSim treats arbitrary search paths in a uniform way, which guarantees the symmetric property of HeteSim. An additional benefit is that HeteSim can evaluate the relatedness of objects with the same- or different types in the same way. Moreover, HeteSim is a semi-metric measure. In other words, HeteSim satisfies the properties of nonnegativity, identity of indiscernibles, and symmetry. It implies that HeteSim can be used in many learning tasks (e.g., clustering and collaborative filtering). We also consider the computation issue of HeteSim and propose four fast computation strategies.

3.1.2 The HeteSim Measure

In many domains, similar objects are more likely to be related to some other similar objects. For example, similar researchers usually publish many similar papers, and similar customers purchase similar commodities. As a consequence, two objects are similar if they are referenced by similar objects. This intuition is also fit for heterogeneous objects. For example, a researcher is more relevant to the conferences that the researcher has published papers in, and a customer is more faithful to the brands that the customer usually purchases. Although the similar idea has been applied in SimRank [6], it is limited to homogeneous networks. When we apply the idea to heterogeneous networks, it faces the following challenges: (1) The relatedness of heterogeneous objects is path-constrained. The meta path not only captures the semantics information but also constrains the walk path. So we need to design a path-based similarity measure. (2) A uniform and symmetric measure should be designed for arbitrary paths. For a given path (symmetric or asymmetric), the measure can evaluate the relatedness of heterogeneous object pair (same or different types) with one single score. In the following section, we will illustrate these challenges and their solutions in detail.

3.1.2.1 Path-Based Relevance Measure

Different from homogeneous networks, the paths in heterogeneous networks have semantics, which makes the relatedness of object pair depend on the given meta path. Following the basic idea that similar objects are related to similar objects, we propose a path-based relevance measure: HeteSim.

Definition 3.1 (*HeteSim*) Given a meta path $P = R_1 \circ R_2 \circ \cdots \circ R_l$, the HeteSim score between two objects s and t ($s \in R_1.S$ and $t \in R_l.T$) is:

$$HeteSim(s, t|R_1 \circ R_2 \circ \cdots \circ R_l) =$$

$$\frac{1}{|O(s|R_1)||I(t|R_l)|} \sum_{i=1}^{|O(s|R_1)|} \sum_{j=1}^{|I(t|R_l)|} HeteSim(O_i(s|R_1), I_j(t|R_l)|R_2 \circ \cdots \circ R_{l-1}) \quad (3.1)$$

where $O(s|R_1)$ is the out-neighbors of s based on relation R_1, and $I(t|R_l)$ is the in-neighbors of t based on relation R_l.

When s does not have any out-neighbors (i.e., $O(s|R_1) = \emptyset$) or t does not have any in-neighbors (i.e., $I(t|R_l) = \emptyset$) following the path, we have no way to infer any relatedness between s and t in this case, so we define their relevance score to be 0. Particularly, we consider objects with the same type to have **self-relation** (denoted as I relation), and each object only has self-relation with itself. It is obvious that an object is just similar to itself for I relation. So its relevance measure can be defined as follows:

Definition 3.2 (*HeteSim based on self-relation*) The HeteSim score between two same-typed objects s and t based on the self-relation I is:

$$HeteSim(s, t|I) = \delta(s, t) \qquad (3.2)$$

where $\delta(s, t) = 1$, if s and t are same, or else $\delta(s, t) = 0$.

Equation 3.1 shows that the computation of $HeteSim(s, t|P)$ needs to iterate over all pairs $(O_i(s|R_1), I_j(t|R_l))$ of (s, t) along the path (s along the path and t against path), and sum up the relatedness of these pairs. Then, we normalize it by the total number of out-neighbors of s and in-neighbors of t. That is, the relatedness between s and t is the average relatedness between the out-neighbors of s and the in-neighbors of t. The process continues until s and t meet along the path. Similar to SimRank [6], HeteSim is also based on pairwise random walk, while it considers the path constraint. As we know, SimRank measures how soon two random surfers are expected to meet at the same node [6]. By contrast, $HeteSim(s, t|P)$ measures how likely s and t will meet at the same node when s follows along the path and t goes against the path.

3.1.2.2   Decomposition of Meta Path

Unfortunately, the source object s and the target object t may not meet along a given path P. For the similarity measure of same-typed objects, the meta paths are usually even-length, even symmetric, so the source object and the target object will meet at the middle objects. However, for the relevance measure of different-typed objects, the meta paths are usually odd-length. In this condition, the source and target objects will never meet at the same objects. Taking the $APVC$ path as an example, authors along the path and conferences against the path will never meet in the same objects. So the original HeteSim is not suitable for odd-length meta paths. In order to solve this difficulty, a basic idea is to transform odd-length paths into even-length paths, and thus, the source and target objects are always able to meet at the same objects. As a consequence, an arbitrary path can be decomposed as two equal-length paths.

When the length l of a meta path $P = (A_1 A_2 \cdots A_{l+1})$ is even, the source objects (along the path) and the target objects (against the path) will meet in the **middle type** object $M = A_{\frac{l}{2}+1}$ on the **middle position** $mid = \frac{l}{2} + 1$, so the meta path P

can be divided into two equal-length path P_L and P_R. That is, $P = P_L P_R$, where $P_L = A_1 A_2 \cdots A_{mid-1} M$ and $P_R = M A_{mid+1} \cdots A_{l+1}$.

When the path length l is odd, the source objects and the target objects will meet at the relation $A_{\frac{l+1}{2}} A_{\frac{l+1}{2}+1}$. In order to let the source and target objects meet at same-typed objects, we can add a middle type object E between the atomic relation $A_{\frac{l+1}{2}} A_{\frac{l+1}{2}+1}$ and maintain the relation between $A_{\frac{l+1}{2}}$ and $A_{\frac{l+1}{2}+1}$ at the same time. Then, the new path becomes $\mathrm{P}' = (A_1 \cdots E \cdots A_{l+1})$ whose length is $l+1$, an even number. The source objects and the target objects will meet in the **middle type** object $M = E$ on the **middle position** $mid = \frac{l+1}{2} + 1$. As a consequence, the new relevance path P' can also be decomposed into two equal-length paths P_L and P_R.

Definition 3.3 (*Decomposition of meta path*) An arbitrary meta path $P = (A_1 A_2 \cdots A_{l+1})$ can be decomposed into two equal-length path P_L and P_R (i.e., $P = P_L P_R$), where $P_L = A_1 A_2 \cdots A_{mid-1} M$ and $P_R = M A_{mid+1} \cdots A_{l+1}$. M and mid are defined as above.

Obviously, for a symmetric path, $P = P_L P_R$, P_R^{-1} is equal to P_L. For example, the meta path $P = APCPA$ can be decomposed as $P_L = APC$ and $P_R = CPA$. For the meta path $APSPVC$, we can add a middle type object E in SP, and thus, the path becomes $APSEPVC$, so $P_L = APSE$ and $P_R = EPVC$.

The next question is how we can add the middle type object E in an atomic relation R between $A_{\frac{l+1}{2}}$ and $A_{\frac{l+1}{2}+1}$. In order to contain original atomic relation, we need to make the R relation be the composition of two new relations. To do so, for each instance of relation R, we can add an instance of E to connect the source and target objects of the relation instance. An example is shown in Fig. 3.1a, where the middle type object E is added in between the atomic relation AB along each path instance.

Definition 3.4 (*Decomposition of atomic relation*) For an atomic relation R, we can add an object type E (called edge object) between the $R.S$ and $R.T$ ($R.S$ and $R.T$ are the source and target object type of the relation R). And thus the atomic relation R is decomposed as R_O and R_I where R_O represents the relation between $R.S$ and E and R_I represents that between E and $R.T$. For each relation instance $r \in R$, an instance $e \in E$ connects $r.S$ and $r.T$. The paths $r.S \to e$ and $e \to r.T$ are the instances of R_O and R_I, respectively.

It is clear that the relation decomposition has the following property, whose proof can be found in [18].

Property 3.1 *An atomic relation R can be decomposed as R_O and R_I, $R = R_O \circ R_I$, and this decomposition is unique.*

Based on this decomposition, the relatedness of two objects with an atomic relation R can be calculated as follows:

Definition 3.5 (*HeteSim based on atomic relation*) The HeteSim score between two different-typed objects s and t based on an atomic relation R ($s \in R.S$ and $t \in R.T$) is:

Fig. 3.1 Decomposition of atomic relation and its HeteSim calculation

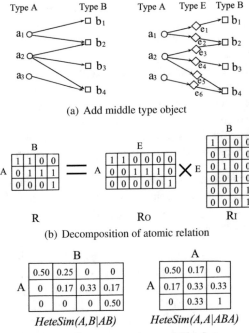

(a) Add middle type object

(b) Decomposition of atomic relation

(c) HeteSim scores before normalization

(d) HeteSim scores after normalization

$$HeteSim(s, t|R) = HeteSim(s, t|R_O \circ R_I) =$$

$$\frac{1}{|O(s|R_O)||I(t|R_I)|} \sum_{i=1}^{|O(s|R_O)|} \sum_{j=1}^{|I(t|R_I)|} \delta(O_i(s|R_O), I_j(t|R_I)) \qquad (3.3)$$

It is easy to find that $HeteSim(s, t|I)$ is a special case of $HeteSim(s, t|R)$, since, for the self-relation I, $I = I_O \circ I_I$ and $|O(s|I_O)| = |I(t|I_I)| = 1$. Definition 3.5 means that HeteSim can measure the relatedness of two different-typed objects with an atomic relation R directly through calculating the average of their mutual influence.

Example 3.1 Figure 3.1a shows an example of decomposition of atomic relation. The relation AB is decomposed into the relations AE and EB. Moreover, the relation AB is the composition of AE and EB as shown in Fig. 3.1b. Two HeteSim examples are illustrated in Fig. 3.1c. We can find that HeteSim justly reflects relatedness of objects. Taking a_2 as example, although a_2 equally connects with b_2, b_3, and b_4, it is more close to b_3, because b_3 only connects with a_2. This information is correctly reflected in the HeteSim score of a_2 based on AB path.

We also find that the similarity of an object and itself is not 1 in HeteSim. Taking the right figure of Fig. 3.1c as example, the relatedness of a_2 and itself is 0.33. It is obviously unreasonable. In the following section, we will normalize the HeteSim and make the relevance measure more reasonable.

3.1.2.3 Normalization of HeteSim

Firstly, we introduce the calculation of HeteSim between any two objects given an arbitrary meta path.

Definition 3.6 (*Transition probability matrix*) For relation $A \xrightarrow{R} B$, W_{AB} is an adjacent matrix between type A and B. U_{AB} is a normalized matrix of W_{AB} along the row vector, which is the transition probability matrix of $A \longrightarrow B$ based on relation R. V_{AB} is a normalized matrix of W_{AB} along the column vector, which is the transition probability matrix of $B \longrightarrow A$ based on relation R^{-1}.

It is easy to prove that the transition probability matrix has the following property. The proof can be found in [18].

Property 3.2 $U_{AB} = V'_{BA}$ and $V_{AB} = U'_{BA}$, where V'_{BA} is the transpose of V_{BA}.

Definition 3.7 (*Reachable probability matrix*) Given a network $G = (V, E)$ following a network schema $S = (A, R)$, a reachable probability matrix PM for a path $P = (A_1 A_2 \cdots A_{l+1})$ is defined as $PM_P = U_{A_1 A_2} U_{A_2 A_3} \cdots U_{A_l A_{l+1}}$ (PM for simplicity). $PM(i, j)$ represents the probability of object $i \in A_1$ reaching object $j \in A_{l+1}$ under the path P.

According to the definition and Property 3.2 of HeteSim, the relevance between objects in A_1 and A_{l+1} based on the meta path $P = A_1 A_2 \cdots A_{l+1}$ is

$$
\begin{aligned}
HeteSim(A_1, A_{l+1}|P) &= HeteSim(A_1, A_{l+1}|P_L P_R) \\
&= U_{A_1 A_2} \cdots U_{A_{mid-1} M} V_{MA_{mid+1}} \cdots V_{A_l A_{l+1}} \\
&= U_{A_1 A_2} \cdots U_{A_{mid-1} M} U'_{A_{mid+1} M} \cdots U'_{A_{l+1} A_l} \qquad (3.4) \\
&= U_{A_1 A_2} \cdots U_{A_{mid-1} M} (U_{A_{l+1} A_l} \cdots U_{A_{mid+1} M})' \\
&= PM_{P_L} PM'_{P_R^{-1}}
\end{aligned}
$$

The above equation shows that the relevance of A_1 and A_{l+1} based on the path P is the inner product of two probability distributions that A_1 reaches the middle type object M along the path and A_{l+1} reaches M against the path. For two instances a and b in A_1 and A_{l+1}, respectively, their relevance based on path P is

$$HeteSim(a, b|P) = PM_{P_L}(a, :)PM'_{P_R^{-1}}(b, :) \qquad (3.5)$$

where $PM_P(a, :)$ means the ath row in PM_P.

We have stated that HeteSim needs to be normalized. It is reasonable that the relatedness of the same objects is 1, so the HeteSim can be normalized as follows:

Definition 3.8 (*Normalization of HeteSim*) The normalized HeteSim score between two objects a and b based on the meta path P is:

$$HeteSim(a, b|P) = \frac{PM_{P_L}(a, :)PM'_{P_R^{-1}}(b, :)}{\sqrt{\|PM_{P_L}(a, :)\|\|PM'_{P_R^{-1}}(b, :)\|}} \qquad (3.6)$$

In fact, the normalized HeteSim is the cosine of the probability distributions of the source object a and target object b reaching the middle type object M. It ranges from 0 to 1. Figure 3.1d shows the normalized HeteSim scores. It is clear that the normalized HeteSim is more reasonable. The normalization is an important step for HeteSim with the following advantages. (1) The normalized HeteSim has nice properties. The following Property 3.4 shows that HeteSim satisfies the identity of indiscernibles. (2) It has a nice interpretation. The normalized HeteSim is the cosine of two vectors representing reachable probability. As Fouss et al. pointed out [3], the angle between the node vectors is a much more predictive measure than the distance between the nodes. In the following section, the HeteSim means the normalized HeteSim.

3.1.2.4   Properties of HeteSim

HeteSim has good properties, which make it useful in many applications. The proof of these properties can be found in [18].

Property 3.3 (Symmetric) $HeteSim(a, b|P) = HeteSim(b, a|P^{-1})$.

Property 3.3 shows the symmetric property of HeteSim. Although PathSim [21] also has the similar symmetric property, it holds only when the path is symmetric and a and b are with the same type. The HeteSim has the more general symmetric property not only for symmetric paths (note that P is equal to P^{-1} for symmetric paths) but also for asymmetric paths.

Property 3.4 (Self-maximum) $HeteSim(a, b|P) \in [0, 1]$. $HeteSim(a, b|P)$ is equal to 1 if and only if $PM_{P_L}(a, :)$ is equal to $PM_{P_R^{-1}}(b, :)$.

Property 3.4 shows HeteSim is well constrained. For a symmetric path P (i.e., $P_L = P_R^{-1}$), $PM_{P_L}(a, :)$ is equal to $PM_{P_R^{-1}}(a, :)$, and thus, $HeteSim(a, a|P)$ is equal to 1. If we define the distance between two objects (i.e., $dis(s, t)$) as $dis(s, t) = 1 - HeteSim(s, t)$, the distance of the same object is zero (i.e., $dis(s, s) = 0$). As a consequence, HeteSim satisfies the identity of indiscernibles. Note that it is a general identity of indiscernibles. For two objects with different types, their HeteSim score is also 1 if they have the same probability distribution on the middle type object. It is reasonable, since they have the similar structure based on the given path.

Since HeteSim obeys the properties of nonnegativity, identity of indiscernibles, and symmetry, we can say that HeteSim is a semi-metric measure [22]. Because of a path-based measure, HeteSim does not obey the triangle inequality. A semi-metric measure has many good merits and can be widely used in many applications [22].

Property 3.5 (Connection to SimRank) *For a bipartite graph $G = (V, E)$ based on the schema $S = (\{A, B\}, \{R\})$, suppose the constant C in SimRank is 1,*

$$SimRank(a_1, a_2) = \lim_{n \rightsquigarrow \infty} \sum_{k=1}^{n} HeteSim(a_1, a_2|(RR^{-1})^k),$$

$$SimRank(b_1, b_2) = \lim_{n \rightsquigarrow \infty} \sum_{k=1}^{n} HeteSim(b_1, b_2|(R^{-1}R)^k).$$

where $a_1, a_2 \in A$, $b_1, b_2 \in B$ and $A \xrightarrow{R} B$. Here HeteSim is the non-normalized version.

This property reveals the connection of SimRank and HeteSim. SimRank sums up the meeting probability of two objects after all possible steps. HeteSim just calculates the meeting probability along the given meta path. If the meta paths explore all possible meta paths among the two types of objects, the sum of HeteSim based on these paths is the SimRank. So we can say that HeteSim is a path-constrained version of SimRank. Through meta paths, HeteSim can subtly evaluate the similarity of heterogeneous objects with fine granularity. This property also implies that HeteSim is more efficient than SimRank, since HeteSim only needs to calculate the meeting probability along the given relevance path, not all possible meta paths.

Moreover, we compare six well-established similarity measures in Table 3.1. There are three similarity measures for heterogeneous networks (i.e., HeteSim, Path-Sim, and PCRW) and three measures for homogeneous networks (i.e., P-PageRank, SimRank, and RoleSim), respectively. Although these similarity measures all evaluate the similarity of nodes by utilizing network structure, they have different properties and features. Three measures for heterogeneous networks all are path-based, since meta paths in heterogeneous networks embody semantics and simplify network structure. Two RW model-based measures (i.e., P-PageRank and PCRW) do not satisfy the symmetric property. Because of satisfying the triangle inequation, RoleSim is a metric, while HeteSim, PathSim, and SimRank are semi-metric. Different from PathSim, which can only measure the similarity of objects with the same type under symmetric paths, the proposed HeteSim can measure the relevance of heterogeneous (same or different-typed) objects under arbitrary (symmetric or asymmetric) paths. Although HeteSim can be considered as a path-constrained extension of SimRank, HeteSim is a general similarity measure in heterogeneous networks with arbitrary schema, not limited to bipartite or N-partite networks.

Table 3.1 Comparison of different similarity measures. Here, RW means random walk, and PRW means pairwise random walk

	Symmetry	Triangle inequation	Path based	Model	Features
HeteSim	√	×	√	PRW	Evaluate relevance of heterogeneous objects based on arbitrary path
PathSim [21]	√	×	√	Path count	Evaluate similarity of same-typed objects based on symmetric path
PCRW [13]	×	×	√	RW	Measure proximity to the query nodes based on given path
SimRank [6]	√	×	×	PRW	Measure similarity of node pairs based on the similarity of their neighbors
RoleSim [9]	√	√	×	PRW	Measure real-valued role similarity based on automorphic equivalence
P-PageRank [7]	×	×	×	RW	Measure personalized views of importance based on linkage structure

3.1.3   Experiments

In the experiments, we validate the effectiveness of the HeteSim on three datasets with three case studies and two learning tasks.

3.1.3.1   Datasets

Three heterogeneous information networks are employed in our experiments.

ACM dataset: The ACM dataset was downloaded from ACM digital library[1] in June 2010. The ACM dataset comes from 14 representative computer science conferences: KDD, SIGMOD, WWW, SIGIR, CIKM, SODA, STOC, SOSP, SPAA, SIGCOMM, MobiCOMM, ICML, COLT, and VLDB. These conferences include 196 corresponding venue proceedings. The dataset has 12 K papers, 17K authors, and 1.8 K author affiliations. After removing stop words in the paper titles and abstracts, we get 1.5 K terms that appear in more than 1% of the papers. The network also includes 73 subjects of these papers in ACM category. The network schema of ACM dataset is shown in Fig. 3.2a. Furthermore, we label the data with the ACM category (i.e., subjects) information. That is, with three major subjects (i.e., H.3, H.2, and C.2), we label 7 conferences, 6772 authors, and 4526 papers.

---

[1] http://dl.acm.org/.

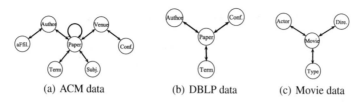

(a) ACM data   (b) DBLP data   (c) Movie data

Fig. 3.2 Network schema of heterogeneous informations

DBLP dataset [8]: The DBLP dataset is a subnetwork collected from DBLP Web site[2] involving major conferences in four research areas: database, data mining, information retrieval, and artificial intelligence, which naturally form four classes. The dataset contains 14 K papers, 20 conferences, 14K authors, and 8.9 K terms, with a total number of 17 K links. In the dataset, 4057 authors, all 20 conferences, and 100 papers are labeled with one of the four research areas. The network schema is shown in Fig. 3.2b.

Movie dataset [17]: The IMDB movie data comes from the Internet Movie Database,[3] which includes movies, actors, directors, and types. A movie heterogeneous network is constructed from the movie data, and its schema is shown in Fig. 3.2c. The movie data contains 1.5 K movies, 5K actors, 551 directors, and 112 types.

3.1.3.2 Case Study

In this section, we demonstrate the traits of HeteSim through case study in three tasks: automatic object profiling, expert finding, and relevance search.

Task 1: Automatic Object Profiling We first study the effectiveness of HeteSim on different-typed relevance measurement in the automatic object profiling task. If we want to know the profile of an object, we can measure the relevance of the object to objects that we are interested in. For example, the academic profile of Christos Faloutsos[4] can be constructed through measuring the relatedness of Christos Faloutsos with related objects, e.g., conferences, affiliations, and other authors. Table 3.2 shows the lists of top relevant objects with various types on ACM dataset. *APVC* path shows the conferences he actively participates. Note that KDD and SIGMOD are the two major conferences Christos Faloutsos participates, which are mentioned in his home page.[5] From the path *APT*, we can obtain his research interests: data mining, pattern discovery, scalable graph mining, and social network. Using *APS* path, we can discover his research areas represented as ACM subjects: database management

---

[2]http://www.informatik.uni-trier.de/~ley/db/.

[3]www.imdb.com/.

[4]http://www.cs.cmu.edu/~christos/.

[5]http://www.cs.cmu.edu/~christos/misc.html.

Table 3.2 Automatic object profiling task on author "Christos Faloutsos" on ACM dataset

Path	APVC		APT		APS		APA	
Rank	Conf.	Score	Terms	Score	Subjects	Score	Authors	Score
1	KDD	0.1198	mining	0.0930	H.2 (database management)	0.1023	Christos Faloutsos	1
2	SIGMOD	0.0284	patterns	0.0926	E.2 (data storage representations)	0.0232	Hanghang Tong	0.4152
3	VLDB	0.0262	scalable	0.0869	G.3 (probability and statistics)	0.0175	Agma Juci M. Traina	0.3250
4	CIKM	0.0083	graphs	0.0816	H.3 (information storage and retrieval)	0.0136	Spiros Papadim-itriou	0.2785
5	WWW	0.0060	social	0.0672	H.1 (models and principles)	0.0135	Caetano Traina, Jr.	0.2680

(H.2) and data storage (E.2). Based on *APA* path, HeteSim finds the most important co-authors, most of which are his Ph.D students.

Task 2: Expert Finding In this case, we want to validate the effectiveness of HeteSim to reflect the relative importance of object pairs through an expert finding task. As we know, the relative importance of object pairs can be revealed through comparing their relatedness. Suppose we know the experts in one domain, the expert finding task here is to find experts in other domains through their relative importance. Table 3.3 shows the relevance scores returned by HeteSim and PCRW on six "conference–author" pairs on ACM dataset. The relatedness of conferences and authors is defined based on the *APVC* and *CVPA* paths which have the same semantics: authors publishing papers in conferences. Due to the symmetric property, HeteSim returns the same value for both paths, while PCRW returns different values for these two paths. Suppose that we are familiar with data mining area and already know that C. Faloutsos is an influential researcher in KDD. Comparing these HeteSim scores, we can find influential researchers in other research areas even if we are not quite familiar with these areas. J.F. Naughton, W.B. Croft, and A. Gupta should be influential researchers in SIGMOD, SIGIR, and SODA, respectively, since they have very similar HeteSim scores to C. Faloutsos. Moreover, we can also deduce that Luo Si and Yan Chen may be active researchers in SIGIR and SIGCOMM, respectively, since they have moderate HeteSim scores. In fact, C. Faloutsos, J.F. Naughton, W.B. Croft, and A. Gupta are top-ranked authors in their research communities. Luo Si and Yan Chen are young professors, and they have done good work in their research areas. However, if the relevance measure is not symmetric (e.g., PCRW), it is very hard to tell which authors are more influential when comparing these relevance scores. For example, the PCRW score of Yan Chen and SIGCOMM is the largest one in the *APVC* path. However, the value is the smallest one for the reversed path (i.e., *CVPA* path).

Table 3.3 Relatedness scores of authors and conferences measured by HeteSim and PCRW on ACM dataset

HeteSim		PCRW			
APVC and *CVPA*		*APVC*		*CVPA*	
Pair	Score	Pair	Score	Pair	Score
C. Faloutsos, KDD	**0.1198**	C. Faloutsos, KDD	0.5517	KDD, C. Faloutsos	**0.0087**
W.B. Croft, SIGIR	**0.1201**	W.B. Croft, SIGIR	0.6481	SIGIR, W.B. Croft	**0.0098**
J.F. Naughton, SIGMOD	**0.1185**	J.F. Naughton, SIGMOD	**0.7647**	SIGMOD, J.F. Naughton	0.0062
A. Gupta, SODA	**0.1225**	A. Gupta, SODA	**0.7647**	SODA, A. Gupta	**0.0090**
Luo Si, SIGIR	0.0734	Luo Si, SIGIR	**0.7059**	SIGIR, Luo Si	0.0030
Yan Chen, SIGCOMM	0.0786	Yan Chen, SIGCOMM	**1**	SIGCOMM, Yan Chen	0.0013

Fig. 3.3 Probability distribution of authors' papers on 14 conferences of ACM dataset

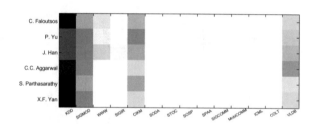

Task 3: Relevance Search based on Path Semantics As we have stated, the path-based relevance measure can capture the semantics of paths. In this relevance search task, we will observe the importance of paths and the effectiveness of semantics capture through the comparison of three path-based measures (i.e., HeteSim, PCRW, and PathSim) and SimRank. This task is to find the top 10 related authors to Christos Faloutsos based on the *APVCVPA* path which means authors publishing papers in same conferences. By ignoring the heterogeneity of objects, we directly run Sim-Rank on whole network and select top ten authors from the rank results which mix different-typed objects together. The comparison results are shown in Table 3.4. At the first sight, we can find that three path-based measures all return researchers having the similar reputation with C. Faloutsos in slightly different orders. However, the results of SimRank are totally against our common sense. We think the reason of bad performances is that SimRank only considers link structure but ignores the link semantics.

In addition, let us analyze the subtle differences of results returned by three path-based measures. The PathSim finds the similar peer authors, such as P. Yu and J. Han. They have the same reputation in data mining field. It is strange for PCRW that the most similar author to C. Faloutsos is not himself, but C. Aggarwal and J. Han. It is

Table 3.4 Top 10 related authors to "Christos Faloutsos" based on *APVCVPA* path on ACM dataset

Rank	HeteSim		PathSim		PCRW		SimRank	
	Author	Score	Author	Score	Author	Score	Author	Score
1	Christos Faloutsos	1	Christos Faloutsos	1	Charu C. Aggarwal	0.0063	Christos Faloutsos	1
2	Srinivasan Parthasarathy	0.9937	Philip Yu	0.9376	Jiawei Han	0.0061	Edoardo Airoldi	0.0789
3	Xifeng Yan	0.9877	Jiawei Han	0.9346	Christos Faloutsos	0.0058	Leejay Wu	0.0767
4	Jian Pei	0.9857	Jian Pei	0.8956	Philip Yu	0.0056	Kensuke Onuma	0.0758
5	Jiong Yang	0.9810	Charu C. Aggarwal	0.7102	Alia I. Abdelmoty	0.0053	Christopher R. Palmer	0.0699
6	Ruoming Jin	0.9758	Jieping Ye	0.6930	Chris B. Jones	0.0053	Anthony Brockwell	0.0668
7	Wei Fan	0.9743	Heikki Mannila	0.6928	Jian Pei	0.0034	Hanghang Tong	0.0658
8	Evimaria Terzi	0.9695	Eamonn Keogh	0.6704	Heikki Mannila	0.0032	Evan Hoke	0.0651
9	Charu C. Aggarwal	0.9668	Ravi Kumar	0.6378	Eamonn Keogh	0.0031	Jia-Yu Pan	0.0650
10	Mohammed J. Zaki	0.9645	Vipin Kumar	0.6362	Mohammed J. Zaki	0.0027	Roberto Santos Filho	0.0648

obviously not reasonable. Our conjecture is that C. Aggarwal and J. Han published many papers in the conferences that C. Faloutsos participated in, so C. Faloutsos has more reachable probability on C. Aggarwal and J. Han than himself along the *APVCVPA* path. HeteSim's results are a little different. The most similar authors are S. Parthasarathy and X. Yan, instead of P. Yu and J. Han. Let us revisit the semantics of the path *APVCVPA*: authors publishing papers in the same conferences. Figure 3.3 shows the reachable probability distribution from authors to conferences along the path *APVC*. It is clear that the probability distribution of papers of S. Parthasarathy and X. Yan on conferences is more close to that of C. Faloutsos, so they should be more similar to C. Faloutsos based on the same conference publication. Although P. Yu and J. Han have the same reputation with C. Faloutsos, their papers are more broadly published in different conferences. So they are not the most similar authors to C. Faloutsos based on the *APVCVPA* path. As a consequence, the HeteSim more accurately captures the semantics of the path.

Since meta path can embody semantics, we can apply HeteSim to do semantic recommendation based on paths given by users. Following this idea, a semantic-based recommended system HeteRecom [17] has been designed.

3.1.3.3   Performance on Query Task

The query task will validate the effectiveness of HeteSim on query search of heterogeneous objects. Since PathSim cannot measure the relatedness of different-typed objects, we only compare HeteSim with PCRW in this experiment. On DBLP dataset, we measure the proximity of conferences and authors based on the *CPA* and *CPAPA* paths. For each conference, we rank its related authors according to their measure scores. Then, we draw the ROC curve of top 100 authors according to the labels of authors (when the labels of author and conference are the same, it is true, else it is false). After that, we calculate the AUC (Area Under ROC Curve) score to evaluate the performances of the ranked results. Note that all conferences and some authors on the DBLP dataset are labeled with one of the four research areas. The larger score means the better performance. We evaluate the performances on 9 representative conferences, and their AUC scores are shown in Table 3.5. We can find that HeteSim consistently outperforms PCRW in most conferences under these two paths. It shows that the proposed HeteSim method can work better than the asymmetric similarity measure PCRW on proximity query task.

3.1.3.4   Performance on Clustering Task

Due to the symmetric property, HeteSim can be applied to clustering tasks directly. In order to evaluate its performance, we compare HeteSim with five well-established similarity measures, including two path-based measures (i.e., PathSim and PCRW) and three homogeneous measures (i.e., SimRank, RoleSim, and P-PageRank). These measures use the same information to determine the pairwise similarity between

Table 3.5 AUC values for the relevance search of conferences and authors based on different paths on DBLP dataset

Paths	Methods	KDD	ICDM	SDM	SIGMOD	VLDB	ICDE	AAAI	IJCAI	SIGIR
CPA	HeteSim	0.811	0.675	0.950	0.766	0.826	0.732	0.811	0.875	0.613
	PCRW	0.803	0.673	0.939	0.758	0.820	0.726	0.806	0.871	0.606
CPAPA	HeteSim	0.845	0.767	0.715	0.831	0.872	0.791	0.817	0.895	0.952
	PCRW	0.844	0.762	0.710	0.822	0.886	0.789	0.807	0.900	0.949

objects. We evaluate the clustering performances on DBLP and ACM datasets. There are three tasks: conference clustering based on *CPAPC* path, author clustering based on *APCPA* path, and paper clustering based on *PAPCPAP* path. For asymmetric measures (i.e., PCRW and P-PageRank), the symmetric similarity matrix can be obtained through the average of similarity matrices based on paths P and P^{-1}. For RoleSim, it is applied in the network constructed by path P. For SimRank and P-PageRank, they are applied in the subnetwork constructed by path P_L (note that the three paths in the experiments are symmetric). Then, we apply normalized cut [16] to perform clustering based on the similarity matrices obtained by different measures. The number of clusters are set as 4 and 3 for DBLP and ACM datasets, respectively. The *NMI* criterion (Normalized Mutual Information) [19] is used to evaluate the clustering performances on conferences, authors, and papers. *NMI* is between 0 and 1 and prothe higher the better. In experiments, the damping factors for P-PageRank, SimRank, and RoleSim are set as 0.9, 0.8, and 0.1, respectively.

The average clustering accuracy results of 100 runs are summarized in Table 3.6. We can find that, on all six tasks, HeteSim achieves best performances on four of them as well as good performances on other two tasks. The mediocre results of PCRW and P-PageRank illustrate that, although symmetric similarity measures can be constructed by the combination of two random walk processes, the simple combination cannot generate good similarity measures. RoleSim aims to detect role similarity, a little bit different from structure similarity, so it has bad performances in these clustering tasks. The experiments show that HeteSim not only does well on similarity measure of same-typed objects but also has the potential as the similarity measure in clustering.

3.1.4 Quick Computation Strategies and Experiments

HeteSim has a high-computation demand for time and space. It is not affordable for online query in large-scale information networks. So a primary strategy is to compute relevance matrix off-line and do online queries with these matrices. For frequently used meta paths, the relatedness matrix $HeteSim(A, B|P)$ can be materialized ahead of time. The online query on $HeteSim(a, B|P)$ will be very fast, since it only needs to locate the row and column in the matrix. However, it also costs much time and space to materialize all frequently used paths. As a consequence, we propose four strategies to fast compute the relevance matrix. Moreover, experiments validate the effectiveness of these strategies.

3.1.4.1 Quick Computation Strategies

The computation of HeteSim includes two phases: matrix multiplication (denoted as MUL, i.e., the computation of PM_{P_L} and $PM_{P_R^{-1}}$) and relevance computation (denoted as REL, i.e., the computation of $PM_{P_L} * PM_{P_R^{-1}}$ and normalization). Through

Table 3.6 Comparison of clustering performances for similarity measures on DBLP and ACM datasets

Methods	DBLP dataset						ACM dataset					
	Venue NMI		Author NMI		Paper NMI		Venue NMI		Author NMI		Paper NMI	
	Mean	Dev.	Mean	Dev.	Mean	Dev.	Mean	Dev.	Mean	Dev.	Mean	Dev.
HeteSim	0.768	0.071	**0.728**	0.083	**0.498**	0.067	**0.843**	0.140	0.405	0.1	**0.439**	0.063
PathSim	0.816	0.078	0.672	0.085	0.383	0.058	0.785	0.164	0.378	0.091	0.432	0.087
PCRW	0.709	0.072	0.710	0.080	0.488	0.039	0.840	0.141	**0.414**	0.092	0.429	0.074
SimRank	**0.888**	0.092	0.685	0.066	0.469	0.031	0.835	0.139	0.375	0.115	0.410	0.073
RoleSim	0.278	0.034	0.501	0.040	0.388	0.049	0.389	0.095	0.293	0.016	0.304	0.017
P-PageRank	0.731	0.086	0.441	0.001	0.421	0.063	0.840	0.164	0.363	0.104	0.407	0.093

analyzing the running time of HeteSim on different phases and paths (the details can be seen in [18]), we find two characteristics of HeteSim computation. (1) The relevance computation is the main time-consuming phase. It implies that the speedup of matrix multiplication may not significantly reduce HeteSim's running time, although this kind of strategies is widely used in accelerating SimRank [6] and PCRW [12]. (2) The dimension and sparsity of matrix greatly affect the efficiency of HeteSim. Although we cannot reduce the running time of relevance computation phase directly, we can accelerate the computation of HeteSim through adjusting matrix dimension and keeping matrix sparse. Based on above idea, we design the following four quick computation strategies.

Dynamic Programming Strategy The matrix multiplication obeys the associative property. Moreover, different computation sequences have different time complexities. The dynamic programming strategy (DP) changes the sequence of matrix multiplication with the associative property. The basic idea of DP is to assign low-dimensioned matrix with the high-computation priority. For a path $P = R_1 \circ R_2 \circ \cdots \circ R_l$, the expected minimal computation complexity of HeteSim can be calculated by the following equation and the computation sequence is recorded by i.

$$Com(R_1 \cdots R_l) =$$
$$\begin{cases} 0 & l = 1 \\ |R_1.S| \times |R_1.T| \times |R_2.T| & l = 2 \\ \arg\min_i \{Com(R_1 \cdots R_i) + Com(R_{i+1} \cdots R_l) + |R_1.S| \times |R_i.T| \times |R_l.T|\} & l > 2 \end{cases}$$
$$(3.7)$$

The above equation can be easily solved by dynamic programming method with the $O(l^2)$ complexity. The running time can be omitted, since l is much smaller than the matrix dimension. Note that the DP strategy only accelerates the MUL phase (i.e., matrix multiplication) and it does not change relevance result, so the DP is an information-lossless strategy.

Truncation Strategy The truncation strategy is based on the hypothesis that removing the probability on those less important nodes would not significantly degrade the performance, which has been proved by many researches [12]. One advantage of this strategy is to keep matrix sparse. The sparse matrix greatly reduces the amount of space and time consumption. The basic idea of truncation strategy is to add a truncation step at each step of random walk. In the truncation step, the relevance value is set with 0 for those nodes when their relevance values are smaller than a threshold ε. A static threshold is usually used in many methods (e.g., Ref. [12]). However, it has the following disadvantage: It may truncate nothing for matrix whose elements all have high probability, and it may truncate most nodes for matrix whose elements all have low probability. Since we usually pay close attention to the top k objects in query task, the threshold ε can be set as the top k relevance value for each search object. For a similarity matrix with size $M \times L$, the k can be dynamically adjusted as follows.

$$k = \begin{cases} L & \text{if } L \leq W \\ \lfloor (L - W)^{\beta} \rfloor + W(\beta \in [0, 1]) & \text{others} \end{cases}$$

where W is the number of top objects, decided by users. The basic idea of dynamic adjustment is that the k slowly increases for super object type (i.e., L is large). The W and β determine the truncation level. The larger W or β will cause the larger k, which means a denser matrix. It is expensive to determine the top k relevance value for each object, so we can estimate the value by the top kM value for the whole matrix. Furthermore, the top kM value can be approximated by the sample data with ratio γ from the raw matrix. The larger γ leads to more accurate approximation with longer running time. In summary, the truncation strategy is an information-loss strategy, which keeps matrix sparse with small sacrifice on accuracy. In addition, it needs additional time to estimate the threshold ε.

Hybrid Strategy As discussed above, the DP strategy can accelerate the MUL phase and the truncation strategy can indirectly speed up the REL phase by keeping sparse matrix. So a hybrid strategy can be designed to combine these two strategies. For the MUL phase, the DP strategy is applied. After obtaining the PM_{P_L} and $PM_{P_R^{-1}}$, the truncation strategy is added. Different from the above truncation strategy, the hybrid strategy only truncates the PM_{P_L} and $PM_{P_R^{-1}}$. The hybrid strategy utilizes the benefits of DP and truncation strategies. It is also an information-loss strategy, since the truncation strategy is employed.

Monte Carlo Strategy Monte Carlo method (MC) is a class of computational algorithms that estimate results through repeating random sampling. It has been applied to compute approximate values of matrix multiplication [2, 12]. In this study, we applied the MC strategy to estimate the value of PM_{P_L} and $PM_{P_R^{-1}}$. The value of $PM_P(a, b)$ can be approximated by the normalized count of the number of times that the walkers visit the node b from a along the path P.

$$PM_P(a, b) = \frac{\#times \ the \ walkers \ visit \ b \ along \ P}{\#walkers \ from \ a}$$

The number of walkers from a (i.e., K) controls the accuracy and amount of computation. The larger K will achieve more accurate estimation with more time cost. An advantage of the MC strategy is that its running time is not affected by the dimension and sparsity of matrix. However, the high-dimension matrix needs larger K for high accuracy. As a sampling method, the MC is also an information-loss strategy.

3.1.4.2   Quick Computation Experiments

We validate the efficiency and effectiveness of quick computation strategies on the ACM dataset. The four paths are used: $(APA)^l$, $(APCPA)^l$, $(APSPA)^l$, and $(TPT)^l$. l means times of path repetition and ranges from 1 to 5. Four quick computation strategies and the original method (i.e., baseline) are employed. The parameters in

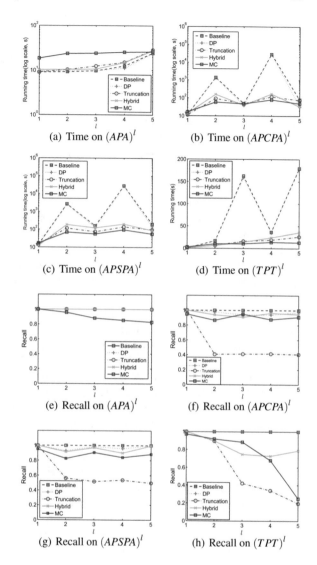

Fig. 3.4 Running time and accuracy of computing HeteSim based on different strategies and paths

truncation process are set as follows: the number of top objects W is 200, β is 0.5, and γ is 0.005. The number of walkers (i.e., K) in MC strategy is 500. The running time and accuracy of all strategies are recorded. In the accuracy evaluation, the relevance matrices obtained by the original method are regarded as the baseline. The accuracy is the *recall* criterion on the top 100 objects obtained by each strategy. All experiments are conducted on machines with Intel Xeon 8-Core CPUs of 2.13 GHz and 64 GB RAM.

Figure 3.4 shows the running time and accuracy of four strategies on different paths. The running time of these strategies is illustrated in Fig. 3.4a–d. We can observe that the DP strategy almost has the same running time with the baseline. It only speeds up the HeteSim computation when the MUL phase dominates the whole running time (e.g., $(APCPA)^5$ and $(APSPA)^5$). It is not the case for the truncation and hybrid strategies, which significantly accelerate the HeteSim computation and have a close speedup ratio on most conditions. Except the APA path, the MC strategy has the highest speedup ratio among all four strategies on most conditions. Then, let us observe their accuracy from Fig. 3.4e–h. The accuracy of the DP strategy is always close to 1. The hybrid strategy achieves the second performances for most paths. The accuracy of the MC strategy is also high for most paths, while it fluctuates on different paths. Obviously, the truncation strategy has the lowest accuracy on most conditions.

As we have noted, the DP, as an information-lossless strategy, only speeds up the MUL phase which is not the main time-consuming part for most paths. So the DP strategy trivially accelerates HeteSim with the accuracy close to 1. The truncation strategy is an information-loss strategy to keep matrix sparse, so it can effectively accelerate HeteSim. That is the reason why the truncation strategy has the high speedup ratio but low accuracy. Because the hybrid strategy combines the benefits of DP and truncation strategy, it has a close speedup ratio to the truncation strategy with higher accuracy. In order to achieve high accuracy, more walkers in the MC strategy are needed for high-dimension or sparse matrix, while the fixed walkers in experiments (i.e., K is 500) make the MC strategy the poor accuracy on some conditions.

According to the analysis above, these strategies are suitable for different paths and scenarios. For very sparse matrix (e.g., $(APA)^1$) and low-dimension matrix (e.g., $(APCPA)^3$), all strategies cannot significantly improve efficiency. However, in these conditions, the HeteSim can be quickly computed without any strategies. For those dense (e.g., $(APCPA)^4$) and high-dimension matrix (e.g., $(APSPA)^4$) which have huge computation overhead, the truncation, hybrid, and MC strategies can effectively improve the HeteSim's efficiency. Particularly, the speedup of the hybrid and MC strategies are up to 100 with little loss in accuracy. If the MUL phase is the main time-consuming part for a path, the DP strategy can also speed up HeteSim greatly without loss in accuracy. The MC strategy has very high efficiency, but its accuracy may degrade for high-dimension matrix. So the appropriate K needs to be set through balancing the efficiency and effectiveness.

3.2 Extension of HeteSim

3.2.1 Overview

Many data mining tasks have been exploited in heterogeneous information network, such as clustering [19] and classification [10]. Among these data mining tasks, similarity measure is a basic and important function, which evaluates the similarity of object pairs on networks. Although similarity measure on homogeneous networks have been extensively studied in the past decades, such as PageRank [15] and SimRank [6], the similarity measure in heterogeneous network is just beginning now and several measures have been proposed including PathSim [21], PCRW [13], and HeteSim [18]. All the three methods are based on meta path [18]. Specially, HeteSim, proposed by Shi et al., has the ability to measure relatedness of objects with the same or different types in a uniform framework. HeteSim has some good properties (e.g., self-maximum and symmetric) and has shown its potential in several data mining tasks. However, we can also find that it has several disadvantages. (1) HeteSim has relatively high computational complexity. Particularly, the adoption of path decomposition approach when it measures the relevance on odd-length path further increases the calculation complexity. (2) Besides, HeteSim cannot be extended to large-scale network with massive data, since its calculation process is based on memory computing. Therefore, it is desired to design a new similarity measure, which not only contains some good properties as HeteSim but also overcomes the disadvantages on computation.

In this chapter, we introduce a new relevance measure method, AvgSim, which is a symmetric measure and uniform measure to evaluate the relevance of same- or different-typed objects. The AvgSim value of two objects is the average of reachable probability under the given path and the reverse path. It guarantees that AvgSim can measure relevance of same or different-typed objects and it has symmetric property. In addition, compared with HeteSim which takes a pairwise random walk, AvgSim does not need to consider the length of path and there is no path decomposition involved. Thus, it is more simple and efficient. Furthermore, we take parallelization of this new algorithm on MapReduce in order to eliminate restriction of memory size and deal with massive data more efficiently in practical applications.

3.2.2 AvgSim: A New Relevance Measure

In this section, we will introduce the new meta path-based relevance measure which is called AvgSim and its definition is as follows.

Definition 3.9 (*AvgSim*) Given a meta path P which is defined on the composite relation $R = R_1 \circ R_2 \circ \ldots \circ R_l$, AvgSim between two objects s and t (s is the source object and t is the target object) is:

$$AvgSim(s, t|P) = \frac{1}{2}[RW(s, t|P) + RW(t, s|P^{-1})] \tag{3.8}$$

$$RW(s, t|R_1 \circ R_2 \circ \ldots \circ R_l) = \frac{1}{|O(s|R_1)|} \sum_{i=1}^{|O(s|R_1)|} RW(O_i(s|R_1), t|R_2 \circ \ldots \circ R_l) \tag{3.9}$$

Equation 3.8 shows the relevance of source object and target object based on meta path P is the arithmetic mean value of random walk result from s to t along P and reversed random walk result from t to s along P^{-1}. Equation 3.9 shows the decomposed step of AvgSim, namely the measure of random walk. The measure takes a random walk step by step from the starting point s to the end point t along path P using an iterative method, where $|O(s|R_1)|$ is the out-neighbors of s based on relation R_1. If there is no out-neighbors of s on R_1, then the relevance value of s and t is 0 because s cannot reach t. We need to calculate the random walk probabilities for each out-neighbor of s to t iteratively, and then, sum them up. Finally, the summation should be normalized by the number of out-neighbors to get the average relatedness.

Then, we will study on how to calculate AvgSim generally with matrix. Given a simple directed meta path $A \xrightarrow{R} B$, where objects A and B are linked though relation R. The relationship between A and B can be expressed by adjacent matrix, denoted as M_{AB}. Two normalized matrices R_{AB} and C_{AB} are generated by normalizing M_{AB} according to row vector and column vector, respectively. R_{AB} and C_{AB} are **transition probability matrix** which represent $A \xrightarrow{R} B$ and $B \xrightarrow{R^{-1}} A$, respectively. According to properties of matrix, we can derive relations $R_{AB} = C'_{BA}$ and $C_{AB} = R'_{BA}$, where R'_{AB} is the transpose of R_{AB}.

If we extend the simple meta path to $P = A_1 \xrightarrow{R_1} A_2 \xrightarrow{R_2} \ldots \xrightarrow{R_l} A_{l+1}$ where R is a composite relation $R = R_1 \circ R_2 \circ \ldots \circ R_l$, then the relationship between A_1 and A_{l+1} is expressed as **reachable probability matrix** which is obtained by multiplying the transition probability matrices along the meta path. The reachable probability matrix of P is defined as $RW_P = R_{A_1A_2}R_{A_2A_3} \ldots R_{A_lA_{l+1}}$, where RW suggests RW_P is the random walk relatedness matrix from object A_1 to A_{l+1} along path P.

Then, we can rewrite AvgSim using the reachable probability matrix according to Eqs. 3.8 and 3.9 as follows.

$$AvgSim(A_1, A_{l+1}|P)$$
$$= \frac{1}{2}[RW(A_1, A_{l+1}|P) + RW(A_{l+1}, A_1|P^{-1})] = \frac{1}{2}[RW_P + RW'_{P-1}] \tag{3.10}$$

According to $C_{AB} = R'_{BA}$, Eq. 3.11 is derived below. We notice that the calculation of AvgSim is unified as two-chain matrix multiplication of transition probability matrices. The only difference between two chains is the normalization form of original adjacent matrix.

$$AvgSim(A_1, A_{l+1}|P) = \frac{1}{2}[R_{A_1A_2}R_{A_2A_3} \ldots R_{A_lA_{l+1}} + (R_{A_{l+1}A_l}R_{A_lA_{l-1}} \ldots R_{A_2A_1})']$$

$$= \frac{1}{2}[R_{A_1A_2}R_{A_2A_3} \ldots R_{A_lA_{l+1}} + C_{A_1A_2}C_{A_2A_3} \ldots C_{A_lA_{l+1}}]$$

$$\tag{3.11}$$

AvgSim can measure the relevance of any heterogeneous or homogeneous object pair based on symmetrical path (e.g., *APCPA*) or asymmetrical path (e.g., *APS*). Besides, the method has a symmetric property, which can be verified easily from the definition equation of AvgSim. However, the calculation of AvgSim mainly lies in the chain matrix multiplication which is time-consuming and restricted of memory size. In order to apply the algorithm in real large-scale heterogeneous information network, we have to consider how to improve the efficiency of AvgSim.

3.2.3 Parallelization of AvgSim

Parallelism [1] is an effective method for processing massive data and improving algorithm's efficiency. According to the features and application scenarios of AvgSim, we parallelize it as the following steps.

1. Since the core calculation of AvgSim is the chain matrix multiplication, we firstly change the order of matrix multiplication operations applying dynamic programming strategy.
2. After Step 1, we turn to focus on large-scale matrix multiplication and it can be parallelized on Hadoop distributed system using MapReduce programming model.

As we know, different orders of operations in chain matrix multiplication leads to different computation time. There exists an optimal order of chain matrix multiplication using dynamic programming, which consumes the shortest computation time. Thus, we can apply dynamic programming to improve the efficiency of parallelized AvgSim.

Parallelization of AvgSim is mainly the parallelization of matrix multiplication after the dynamic programming process. Here, we use the "block matrix multiplication" method on MapReduce to transform multiplication of two large matrices into several multiplications of smaller matrices. This method is flexible for selecting dimensions of block matrix according to the configuration of Hadoop cluster, and it avoids exceeding the memory size. The parallelization of block matrix multiplication is implemented by two-round MapReduce computing. The detailed algorithms can be found in [14].

Applying two-round MapReduce algorithm above iteratively to the chain matrix multiplication which is reordered by dynamic programming, we can obtain one of the two reachable probability matrices of AvgSim (e.g., RW_P, which is measured in the given meta path P), and the other probability matrix (RW'_{P-1}) can be obtained in the same procedure. Finally, the relevance matrix is derived by taking the arithmetic mean of these two reachable probability matrices.

3.2.4   Experiments

Three datasets, ACM dataset, DBLP dataset, and Matrix dataset, are used in experiments. In detail, the ACM dataset contains 17 K authors, 1.8K author affiliations, 12K papers, and 14 computer science conferences including 196 corresponding venue proceedings. We also extract 1.5 K terms and 73 subjects from these papers. The DBLP dataset contains 14 K papers, 14K authors, 20 conferences, and 8.9 K terms. And we label 20 conferences, 100 papers, and 4057 authors in the dataset with four research areas including database, data mining, information retrieval, and artificial intelligence for experiment use. And the matrix dataset *(40 matrices in total)* contains several artificially generated large-scale sparse square matrices, whose dimensions are $1000 \times 1000, 5000 \times 5000, 10,000 \times 10,000, 20,000 \times 20,000, 40,000 \times 40,000, 80,000 \times 80,000, 100,000 \times 100,000$, and $150,000 \times 150,000$, respectively. And the sparsity of each matrix is 0.0001, 0.0003, 0.0005, 0.0007, and 0.001.

3.2.4.1   Effectiveness of AvgSim

In this section, we design experiments to validate the effectiveness and efficiency of AvgSim. We design two tasks to verify the effectiveness of AvgSim, which are query task and clustering task, respectively.

In the query task, we compare the performance of AvgSim with both HeteSim and PCRW though measuring the relevance of heterogeneous objects on DBLP dataset. Based on labels of the dataset, we calculate the AUC score to evaluate the performances of different methods, where the query task is to find authors for each conference based on the path *CPA*. We evaluated 9 out of 20 marked conferences, whose AUC values are shown in Table 3.7. We notice that AvgSim gets the highest value on 8 conferences, which means AvgSim performs better than other two methods in this query task.

In the clustering task, we compare the performance of AvgSim with both HeteSim and PathSim through measuring the similarity of homogeneous objects on DBLP dataset. We firstly apply three algorithms, respectively, to derive the similarity matrices on three meta paths including *CPAPC*, *APCPA*, and *PAPCPAP*. We perform clustering task based on the similarity matrices with normalized cut and

Table 3.7 AUC values for relevance search of conferences and authors based on CPA path on DBLP dataset

	KDD	ICDM	SDM	SIGMOD	VLDB	ICDE	AAAI	IJCAI	SIGIR
HeteSim	0.8111	0.6752	**0.6132**	0.7662	0.8262	**0.7322**	0.8110	0.8754	0.9504
PCRW	0.8030	0.6731	0.6068	0.7588	0.8200	0.7263	0.8067	0.8712	0.9390
AvgSim	**0.8117**	**0.6753**	0.6072	**0.7668**	**0.8274**	0.7286	**0.8114**	**0.8764**	**0.9525**

Table 3.8 Clustering accuracy results for path-based relevance measures on DBLP dataset

	Venue NMI	Author NMI	Paper NMI
PathSim	0.8162	0.6725	0.3833
HeteSim	0.7683	0.7288	0.4989
AvgSim	**0.8977**	**0.7556**	**0.5101**

then evaluate the performances on conferences, authors, and papers using *NMI* criterion (Normalized Mutual Information). The clustering accuracy result is shown in Table 3.8, and AvgSim obtains the highest *NMI* value in all the three tasks. In all, the results of query task and clustering task suggest that AvgSim performs well in effectiveness.

3.2.4.2 Efficiency of AvgSim

In this section, we will verify the efficiency of AvgSim on ACM dataset. We take relevance measure experiments of AvgSim and HeteSim, respectively, based on meta paths including $(APCPA)^l$ and $(TPT)^l$, where l is the number of path repetitions with a range from 1 to 5.

Figure 3.5a, b shows the relationship between running time and different meta paths for each method. We notice that the running time of HeteSim exhibits great fluctuations with the change of path length, while AvgSim is much stable. According to the definition of AvgSim, the longer paths (i.e., l) it measures, the more matrices need to be multiplied, and thus, the running time increases persistently. In contrast, the calculation of HeteSim needs two steps including matrix multiplication and relevance computation. In the matrix multiplication step, HeteSim calculates the reachable probability matrices from source and target nodes to the middle node, respectively. The longer paths it measures, the more time it needs. In relevance computation step, the relevance matrix is the multiplication of two probability matrices in previous step. The time for the second step is determined by the scale of middle node. In all, the relevance computation of HeteSim affects its performance to a great extent and it will be relatively poor for large-scale matrices. Conversely, AvgSim performs much more stable, and its efficiency is only related to the matrix dimension and

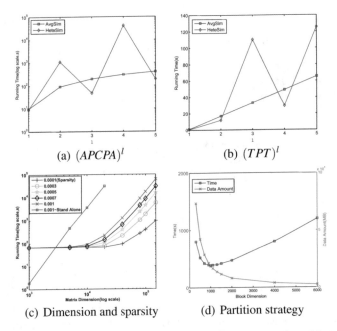

Fig. 3.5 Running time of AvgSim and HeteSim based on different meta paths and factors affecting parallelized block matrix multiplication: **a** Running time on $(APCPA)^l$; **b** Running time on $(TPT)^l$; **c** Matrix dimension and sparsity factors; **d** Partition strategy factor

meta path length, which can be improved by the parallelized matrix multiplication on MapReduce.

All parallelized matrix multiplication experiments are conducted in a cluster composed of 7 machines with 4-cores E3-1220 V2 CPUs of 3.10 GHz and 32 GB RAM running on RedHat 4 operating system. The experiments will study several factors affecting block matrix multiplication, including matrix, matrix sparsity, and partition strategy (i.e., dimensions of blocks). Results will reflect the performance of parallelized AvgSim algorithm.

Figure 3.5c shows the effect of matrix dimensions and matrix sparsity on the running time of parallelized block matrix multiplication. All the matrix multiplications are done on the *Matrix* dataset, and it applies the partition strategy of 1000 × 1000 block matrix. We notice from Fig. 3.5c that the larger dimensions or more density of matrix are, the more time in matrix multiplication is required. And the comparison results between stand-alone and parallelized matrix multiplication with the sparsity of 0.001 shows that the stand-alone algorithm costs shorter time for a quite small matrix dimension because the parallelized algorithm spends lots of time in the starting task nodes of Hadoop cluster and resources of cluster are not fully utilized for a small amount of calculations. However, the efficiency of parallelized algorithm is much better as the matrix dimension increases. Besides, the stand-alone algorithm is restricted of memory size, so there are no results derived in the last three large-scale matrix multiplications shown in Fig. 3.5c.

Figure 3.5d shows the effect of intermediate data amount and partition strategy of block matrix multiplication. There are 11 kinds of partition strategies with the square block matrix dimensions from 300×300 to 6000×6000, where the square matrix is with the dimension of $100,000 \times 100,000$ and the sparsity of 0.0001 in the experiment. We notice from Fig. 3.5d that the intermediate data amount of matrix multiplication decreases gradually with the increase of block dimension. In contrast, the running time reaches its minimum value at 5th data point. Smaller intermediate data amount results in less disk IO operations and data amount transmitted by shuffle, which also leads to shorter time and better performance to a certain extent as the data points before 1000 near 1000 reflected. However, the excessive large block dimension will reduce the concurrent granularity and increase the amount of calculations for single node, which conversely results in longer time of computation as the data points after 1000 reflected.

In all, the appropriate partition strategy and sufficient sizes of cluster greatly affect the efficiency in parallelized block matrix multiplications. Applying parallelization method, AvgSim gains the ability to measure relevance in larger-scale networks with massive data efficiently.

3.3 Conclusion

In this chapter, we study the relevance search problem which measures the relatedness of heterogeneous objects in heterogeneous networks. We introduce a general relevance measure, called HeteSim. As a path-constraint and semi-metric measure, HeteSim can measure the relatedness of same-typed or different-typed objects in a uniform framework. In addition, we also present a modification of HeteSim. Extensive experiments validate the effectiveness and efficiency of the proposed measures on evaluating the relatedness of heterogeneous objects.

The similarity measure of objects in heterogeneous networks is an important and basic task, which can be used in many applications. There are some interesting directions for future work. Similarity measures are designed for more complex HIN, such as hybrid network integrating heterogeneous features and text information, and multiple or weighted meta paths. In addition, similarity measures are widely used in real applications where the network scales are usually huge. We need to design the efficient and parallelized computation methods.

References

1. Cao, L., Cho, B., Kim, H.D., Li, Z., Tsai, M.H., Gupta, I.: Delta-simrank computing on mapre-duce. In: Big Data Workshop, pp. 28–35 (2012)
2. Fogaras, D., Rácz, B., Csalogány, K., Sarlós, T.: Towards scaling fully personalized PageRank: algorithms, lower bounds, and experiments. Internet Math. **2**(3), 333–358 (2005)
3. Fouss, F., Pirotte, A., Renders, J.M., Saerens, M.: Random-walk computation of similarities between nodes of a graph with application to collaborative recommendation. IEEE Trans. Knowl. Data Eng. **19**(3), 355–369 (2007)
4. Han, J.: Mining heterogeneous information networks by exploring the power of links. In: DS, pp. 13–30 (2009)
5. Jamali, M., Lakshmanan, L.V.S.: HeteroMF: recommendation in heterogeneous information networks using context dependent factor models. In: WWW, pp. 643–654 (2013)
6. Jeh, G., Widom, J.: SimRank: A measure of structural-context similarity. In: KDD, pp. 538–543 (2002)
7. Jeh, G., Widom, J.: Scaling personalized web search. In: WWW, pp. 271–279 (2003)
8. Ji, M., Sun, Y., Danilevsky, M., Han, J., Gao, J.: Graph regularized transductive classification on heterogeneous information networks. In: ECML/PKDD, pp. 570–586 (2010)
9. Jin, R., Lee, V.E., Hong, H.: Axiomatic ranking of network role similarity. In: KDD, pp. 922–930 (2011)
10. Kong, X., Yu, P.S., Ding, Y., Wild, D.J.: Meta path-based collective classification in heterogeneous information networks. In: CIKM, pp. 1567–1571 (2012)
11. Konstan, J.A., Miller, B.N., Maltz, D., Herlocker, J.L., Gordon, L.R., Riedl, J.: GroupLens: applying collaborative filtering to Usenet news. Commun. ACM **40**(3), 77–87 (1997)
12. Lao, N., Cohen, W.: Fast query execution for retrieval models based on path constrained random walks. In: KDD, pp. 881–888 (2010)
13. Lao, N., Cohen, W.W.: Relational retrieval using a combination of path-constrained random walks. Mach. Learn. **81**(2), 53–67 (2010)
14. Meng, X., Shi, C., Li, Y., Zhang, L., Wu, B.: Relevance measure in large-scale heterogeneous networks. In: APWeb, pp. 636–643 (2014)
15. Page, L., Brin, S., Motwani, R., Winograd, T.: The pagerank citation ranking: bringing order to the web. In: Stanford InfoLab, pp. 1–14 (1998)
16. Shi, J., Malik, J.: Normalized cuts and image segmentation. IEEE Trans. Pattern Anal. Mach. Intell. **22**(8), 888–905 (2000)
17. Shi, C., Zhou, C., Kong, X., Yu, P.S., Liu, G., Wang, B.: HeteRecom: a semantic-based recommendation system in heterogeneous networks. In: KDD, pp. 1552–1555 (2012)
18. Shi, C., Kong, X., Huang, Y., Philip, S.Y., Wu, B.: Hetesim: a general framework for relevance measure in heterogeneous networks. IEEE Trans. Knowl. Data Eng. **26**(10), 2479–2492 (2014)
19. Sun, Y., Han, J., Zhao, P., Yin, Z., Cheng, H., Wu, T.: RankClus: integrating clustering with ranking for heterogeneous information network analysis. In: EDBT, pp. 565–576 (2009)
20. Sun, Y., Yu, Y., Han, J.: Ranking-based clustering of heterogeneous information networks with star network schema. In: KDD, pp. 797–806 (2009)
21. Sun, Y.Z., Han, J.W., Yan, X.F., Yu, P.S., Wu, T.: PathSim: Meta path-based Top-K similarity search in heterogeneous information networks. In: VLDB, pp. 992–1003 (2011)
22. Xia, Q.: The geodesic problem in quasimetric spaces. J. Geom. Anal. **19**(2), 452–479 (2009)
23. Zhu, J., De Vries, A.P., Demartini, G., Iofciu, T.: Evaluating relation retrieval for entities and experts. In: Proceedings of the SIGIR 2008 Workshop on Future Challenges in Expertise Retrieval (fCHER), pp. 41–44 (2008)

Chapter 4
Path-Based Ranking and Clustering

Abstract As newly emerging network models, heterogeneous information networks have many unique features, e.g., complex structures and rich semantics. Moreover, meta path, the sequence of relations connecting two object types, is an effective tool to integrate different types of objects and mine the semantic information in this kind of networks. The unique characteristics of meta path make the data mining on heterogeneous network more interesting and challenging. In this chapter, we will introduce two basic data mining tasks, ranking and clustering, on heterogeneous information network. Furthermore, we introduce the HRank method to evaluate the importance of multiple types of objects and meta paths, and present the HeProjI algorithm to solve the heterogeneous network projection and integration of clustering and ranking tasks.

4.1 Meta Path-Based Ranking

4.1.1 Overview

It is an important research problem to evaluate object importance or popularity, which can be used in many data mining tasks. Many methods have been developed to evaluate object importance, such as PageRank [13], HITS [7], and SimRank [5]. In these literatures, objects ranking is done in a homogeneous network in which objects or relations are the same. For example, both PageRank and HITS rank the web pages in WWW.

However, in many real network data, there are many different types of objects and relations, which can be organized as heterogeneous networks. Formally, heterogeneous information networks (HIN) are the logical networks involving multiple types of objects as well as multiple types of links denoting different relations [4]. Recently, many data mining tasks have been exploited in this kind of networks,

© Springer International Publishing AG 2017
C. Shi and P.S. Yu, *Heterogeneous Information Network Analysis and Applications*, Data Analytics, DOI 10.1007/978-3-319-56212-4_4

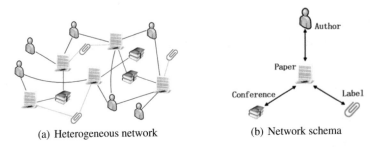

(a) Heterogeneous network   (b) Network schema

Fig. 4.1 A heterogeneous information network example on bibliographic data. **a** shows heterogeneous objects and their relations. **b** shows the network schema

such as similarity measure [14, 25], clustering [23], and classification [6], among which ranking is an important but seldom exploited task.

Figure 4.1a shows an HIN example in bibliographic data, and Fig. 4.1b illustrates its network schema which depicts object types and their relations. In this example, it contains objects from four types of objects: papers (P), authors (A), labels (L, categories of papers), and conferences (C). There are links connecting different types of objects. The link types are defined by the relation between the two object types. In this network, several interesting, yet seldom exploited, ranking problems can be proposed.

- One may be interested in the importance of one type of objects and ask the following questions:
  Q. 1.1 Who are the most influential authors?
  Q. 1.2 Who are the most influential authors in data mining field?
- As we know, some object types have an effect on each other. For example, influential authors usually publish papers in reputable conferences. So one may pay attention to the importance of multiple types of objects simultaneously and ask the following questions:
  Q. 2.1 Who are the most influential authors and which reputable conferences did those influential authors publish their papers on?
  Q. 2.2 Who are the most influential authors and which reputable conferences did those influential authors publish their papers on in data mining field?
- Furthermore, one may wonder which factor mostly affects the importance of objects, since the importance of objects is affected by many factors. So he may ask the questions like this:
  Q. 3 Who are the most influential authors and which factors make those most influential authors be most influential?

Although the ranking problem in homogeneous networks has been well studied, the above ranking problems are unique in HIN (especially $Q. 2$ and $Q. 3$), which are seldom studied until now. Since there are multiple types of objects in HIN, it is possible to analyze the importance of multiple types of objects (i.e., $Q. 2$) as well as affecting factors (i.e., $Q. 3$) together.

In this chapter, we study the ranking problem in HIN and propose a ranking method, HRank, to evaluate the importance of multiple types of objects and meta paths in HIN. For *Q. 1* and *Q. 2*, a path-based random walk model is proposed to evaluate the importance of single or multiple types of objects. The different meta paths connecting two types (same or different types) of objects have different semantics and transitive probability, and thus lead to different random walk processes and ranking results. Although meta path has been widely used to capture the semantics in HIN [14, 25], it coarsely depicts object relations. By employing the meta path, we can answer the *Q. 1.1* and *Q. 2.1*, but cannot answer the *Q. 1.2* and *Q. 2.2*. In order to overcome the shortcoming existing in meta path, we propose the *constrained meta path* concept, which can effectively describe this kind of subtle semantics. The constrained meta path assigns constraint conditions on meta path. Through adopting the constrained meta path, we can answer the *Q. 1.2* and *Q. 2.2*.

Moreover, in HIN, based on different paths, the objects have different ranking values. The comprehensive importance of objects should consider all kinds of factors (the factors can be embodied by constrained meta paths), which have different contribution to the importance of objects. In order to evaluate the importance of objects and meta paths simultaneously (i.e., answer *Q. 3*), we further propose a co-ranking method which organizes the relation matrices of objects on different constrained meta paths as a tensor. A random walk process is designed on this tensor to co-rank the importance of objects and paths simultaneously. That is, random walkers surf in the tensor, where the stationary visiting probability of objects and meta paths is considered as the HRank score of objects and paths.

4.1.2 The HRank Method

Since the importance of objects is related to the meta path designated by users, we propose the path-based ranking method HRank in heterogeneous networks. In order to answer the three ranking problems proposed above, we design three versions of HRank, respectively.

4.1.2.1 Constrained Meta Path

As an effective semantic capturing method, the meta path has been widely used in many data mining tasks in HIN, such as similarity measure [14, 25], clustering [23], and classification [8]. However, meta path may fail to capture subtle semantics in some situations. Taking Fig. 4.1b as an example, the *APA* path cannot reveal the co-author relations in a certain research field, such as data mining and information retrieval. Although Jiawei Han has co-worked many papers with Philip S. Yu in the data mining field, they never co-work in the operation system field. The *APA* path cannot subtly reflect this difference.

In order to overcome the shortcomings in meta path, we propose the concept of constrained meta path, defined as follows.

Definition 4.1 (*Constrained meta path*) A constrained meta path is a meta path based on a certain constraint which is denoted as $\mathrm{CP} = P|\mathrm{C}$. $P = (A_1 A_2 \ldots A_l)$ is a meta path, while C represents the constraint on the objects in the meta path.

Note that the C can be one or multiple constraint conditions on objects. Taking Fig. 4.1b as an example, the constrained meta path $APA|P.L = \text{“DM”}$ represents the co-author relations of authors in data mining field through constraining the label of papers with data mining (DM). Similarly, the constrained meta path $APCPA|P.L = \text{“DM”}\&\&C = \text{“CIKM”}$ represents the co-author relations of authors in CIKM conference, and the papers of authors are in data mining field. Obviously, compared to meta path, the constrained meta path conveys richer semantics by subdividing meta paths under distinct conditions. Particularly, when the length of meta path is 1 (i.e., a relation), the constrained meta path degrades to a **constrained relation**. In other words, the constrained relation confines constraint conditions on objects of the relation.

For a relation $A \xrightarrow{R} B$, we can obtain its transition probability matrix as follows.

Definition 4.2 (*Transition probability matrix*) W_{AB} is an adjacent matrix between type A and B on relation $A \xrightarrow{R} B$. U_{AB} is the normalized matrix of W_{AB} along the row vector, which is the transition probability matrix of $A \xrightarrow{R} B$.

Then, we make some constraints on objects of the relation $A \xrightarrow{R} B$ (i.e., constrained relation). We can have the following definition.

Definition 4.3 (*Constrained transition probability matrix*) W_{AB} is an adjacent matrix between type A and B on relation $A \xrightarrow{R} B$. Suppose there is a constraint C on object type A. The constrained transition probability matrix U'_{AB} of constrained relation $R|\mathrm{C}$ is $U'_{AB} = M_C U_{AB}$, where M_C is the constraint matrix generated by the constraint condition C on object type A.

The constraint matrix M_C is usually a diagonal matrix whose dimension is the number of objects in object type A. The element in the diagonal is 1 if the corresponding object satisfies the constraint, else the element in the diagonal is 0. For example, in the path $PC|C = \text{“CIKM”}$, M_C is a diagonal matrix of conferences, where the "*CIKM*" column is 1 and the others are 0. Similarly, we can confine the constraint on the object type B or both types. Note that the transition probability matrix is a special case of the constrained transition probability matrix, when we let the constraint matrix M_C be the identity matrix I.

Given a network $G = (V, E)$ following a network schema $S = (\mathrm{A}, \mathrm{R})$, we can define the meta path-based reachable probability matrix as follows.

Definition 4.4 (*Meta path-based reachable probability matrix*) For a meta path $P = (A_1 A_2 \cdots A_{l+1})$, the meta path-based reachable probability matrix PM is defined as $PM_P = U_{A_1 A_2} U_{A_2 A_3} \cdots U_{A_l A_{l+1}}$. $PM_P(i, j)$ represents the probability of object $i \in A_1$ reaching object $j \in A_{l+1}$ under the path P.

Similarly, we have the following definition for constrained meta path.

Definition 4.5 (*Constrained meta path-based reachable probability matrix*) For a constrained meta path $\text{CP} = (A_1 A_2 \cdots A_{l+1} | C)$, the constrained meta path-based reachable probability matrix is defined as $PM_{\text{CP}} = U'_{A_1 A_2} U'_{A_2 A_3} \cdots U'_{A_l A_{l+1}}$. $PM_{\text{CP}}(i, j)$ represents the probability of object $i \in A_1$ reaching object $j \in A_{l+1}$ under the constrained meta path $P|C$.

In fact, if there is no constraint on the objects of a relation $A_i \xrightarrow{R} A_{i+1}$, $U'_{A_i A_{i+1}}$ is equal to $U_{A_i A_{i+1}}$. If there is a constraint on the objects, we only consider the objects that satisfy the constraint. For simplicity, we use the reachable probability matrix and the M_P to represent the constrained meta path-based reachable probability matrix in the following section.

4.1.2.2 Ranking Based on Symmetric Meta Paths

In order to evaluate the importance of one type of objects (i.e., *Q. 1*), we design the HRank-SY method based on symmetric constrained meta paths, since the constrained meta paths connecting one type of objects are usually symmetric, such as $APA|P.L = $ "*DM*".

For a symmetric constrained meta path $P = (A_1 A_2 \ldots A_l | C)$, P is equal to P^{-1} and A_1 and A_l are the same. Similar to PageRank [13], the importance evaluation of object A_1 (i.e., A_l) can be considered as a random walk process in which random walkers wander from type A_1 to type A_l along the path P. The HRank value of object A_1 (i.e., $R(A_1|P)$) is the stable visiting probability of random walkers, which is defined as follows:

$$R(A_1|P) = \alpha R(A_1|P)M_P + (1 - \alpha)E \tag{4.1}$$

where M_P is the constrained meta path-based reachable probability matrix as defined above. E is the restart probability vector for convergence. It is set equally for all objects of type A_1, which is $1/|A_1|$. α is the decay factor, which can be set with 0.85 as the parameter experiments suggested. HRank-SY and PageRank both have the same idea that the importance of objects is decided by the visiting probability of random surfers. Different from PageRank, the random surfers in HRank-SY should wander along the constrained meta path to visit objects.

As shown in Fig. 4.2, the red broken line illustrates an example of the process of calculating rank values, where the CP is $APA|P.L = $ "*DM*". The concrete calculating process is as follows:

Fig. 4.2 An example of the computation process of HRank. The *blue* and *red broken lines* represent the process on the symmetric and asymmetric constrained meta path, respectively

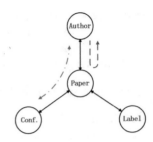

$$R(Author|\text{CP}) = \alpha R(Author|\text{CP})M_{\text{CP}} + (1 - \alpha)E$$
$$M_{\text{CP}} = U'_{AP}U'_{PA} = U_{AP}M_PM_PU_{PA}$$
(4.2)

where M_P is the constraint matrix on object type P (paper).

4.1.2.3 Ranking Based on Asymmetric Meta Paths

For the question *Q. 2*, we propose the HRank-AS method based on asymmetric constrained meta paths, since the paths connecting different types of objects are asymmetric. For an asymmetric constrained meta path $P = (A_1A_2 \ldots A_l|C)$, P is not equal to P^{-1}. Note that A_1 and A_l are either of the same or different types, such as $APC|P.L = \text{“}DM\text{”}$ and $PCPLP|C = \text{“}CIKM\text{”}$.

Similarly, HRank-AS is also based on a random walk process that random walkers wander between A_1 and A_l along the path. The ranks of A_1 and A_l can be seen as the visiting probability of walkers, which are defined as follows:

$$R(A_l|P^{-1}) = \alpha R(A_1|P)M_P + (1 - \alpha)E_{A_l}$$
$$R(A_1|P) = \alpha R(A_l|P^{-1})M_{P^{-1}} + (1 - \alpha)E_{A_1}$$
(4.3)

where M_P and $M_{P^{-1}}$ are the reachable probability matrix of path P and P^{-1}. E_{A_1} and E_{A_l} are the restart probability of A_1 and A_l. Obviously, HRank-SY is the special case of HRank-AS. When the path P is symmetric, Eq. 4.3 is the same with Eq. 4.1.

The blue broken line in Fig. 4.2 illustrates an example which simultaneously evaluates the importance of authors and conferences. Here, the CP is $APC|P.L = \text{“}DM\text{”}$. The concrete calculating process is as follows:

$$R(Conf.|\text{CP}) = \alpha R(Aut.|\text{CP})M_{\text{CP}} + (1 - \alpha)E_{Conf.}$$
$$R(Aut.|\text{CP}) = \alpha R(Conf.|\text{CP})M_{\text{CP}^{-1}} + (1 - \alpha)E_{Aut.}$$
$$M_{\text{CP}} = U'_{AP}U'_{PC} = U_{AP}M_PM_PU_{PC}$$
$$M_{\text{CP}^{-1}} = U'_{CP}U'_{PA} = U_{CP}M_PM_PU_{PA}$$
(4.4)

where M_P is the constraint matrix on object type P (paper).

4.1.2.4 Co-ranking for Objects and Relations

Until now, we have created methods to rank same or different types of objects under a certain constrained meta path. However, there are many constrained meta paths in heterogeneous networks. It is an important issue to automatically determine the importance of paths [23, 25], since it is usually hard for us to identify which relation is more important in real applications. To solve this problem (i.e., $Q.$ 3), we propose the HRank-CO to co-rank the importance of objects and relations. The basic idea is based on an intuition that important objects are connected to many other objects through a number of important relations and important relations connect many important objects. So we organize the multiple relation networks with a tensor, and a random walk process is designed on this tensor. The method not only can comprehensively evaluate the importance of objects by considering all constrained meta paths, but also can rank the contribution of different constrained meta paths.

In Fig. 4.3a, we show an example of multiple relations among objects, generated by multiple meta paths. There are three objects of type A, three objects of type B, and three types of relations among them. These relations are generated by three constrained meta paths with type A as the source type and type B as the target type. To describe the multiple relations among objects, we use the representation of tensor which is a multidimensional array. We call $X = (x_{i,j,k})$ a third-order tensor, where $x_{i,j,k} \in R$, for $i = 1, \cdots, m, j = 1, \cdots, l, k = 1, \cdots, n$. $x_{i,j,k}$ represents the times that object i is related to object k through the jth constrained meta path. For example, Fig. 4.3b is a three-way array, where each two-dimensional slice represents an adjacency matrix for a single relation. So the data can be represented as a tensor of size $3 \times 3 \times 3$. In the multirelational network, we define the transition probability tensor to present the transition probability among objects and relations.

Definition 4.6 (*Transition probability tensor*) In a multirelational network, X is the tensor representing the network. F is the normalized tensor of X along the column vector. R is the normalized tensor of X along the tube vector. T is the normalized

Fig. 4.3 An example of multirelations of objects generated by multiple paths: **a** the graph representation; **b** the corresponding tensor representation

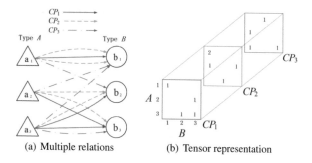

(a) Multiple relations   (b) Tensor representation

tensor of X along the row vector. F, R, and T are called the transition probability
tensors which can be denoted as follows:

$$
\begin{aligned}
f_{i,j,k} &= \frac{x_{i,j,k}}{\sum_{i=1}^{m} x_{i,j,k}} \quad i = 1, 2, \ldots, m \\
r_{i,j,k} &= \frac{x_{i,j,k}}{\sum_{j=1}^{l} x_{i,j,k}} \quad j = 1, 2, \ldots, l \\
t_{i,j,k} &= \frac{x_{i,j,k}}{\sum_{k=1}^{n} x_{i,j,k}} \quad k = 1, 2, \ldots, n
\end{aligned}
\tag{4.5}
$$

$f_{i,j,k}$ can be interpreted as the probability of object i (of type A) being the visiting
object when relation j is used and the current object being visited is object k (of type
B), $r_{i,j,k}$ represents the probability of using relation j given that object k is visited
from object i, and $t_{i,j,k}$ can be interpreted as the probability of object k being visited,
given that object i is currently the visiting object and relation j is used. The meaning
of these three tensors can be defined formally as follows:

$$
\begin{aligned}
f_{i,j,k} &= Prob(X_t = i | Y_t = j, Z_t = k) \\
r_{i,j,k} &= Prob(Y_t = j | X_t = i, Z_t = k) \\
t_{i,j,k} &= Prob(Z_t = k | X_t = i, Y_t = j)
\end{aligned}
\tag{4.6}
$$

in which X_t, Z_t, and Y_t are three random variables representing visiting at certain
object of type A or type B and using certain relation respectively at the time t.

Now, we define the stationary distributions of objects and relations as follows:

$$
\begin{aligned}
x &= (x_1, x_2, \cdots, x_m)^T \\
y &= (y_1, y_2, \cdots, y_l)^T \\
z &= (z_1, z_2, \cdots, z_n)^T
\end{aligned}
\tag{4.7}
$$

in which

$$
\begin{aligned}
x_i &= \lim_{t \to \infty} Prob(X_t = i) \\
y_j &= \lim_{t \to \infty} Prob(Y_t = j) \\
z_k &= \lim_{t \to \infty} Prob(Z_t = k)
\end{aligned}
\tag{4.8}
$$

From the above equations, we can get:

$$Prob(X_t = i) = \sum_{j=1}^{l}\sum_{k=1}^{n} f_{i,j,k} \times Prob(Y_t = j, Z_t = k)$$

$$Prob(Y_t = j) = \sum_{i=1}^{m}\sum_{k=1}^{n} r_{i,j,k} \times Prob(X_t = i, Z_t = k) \qquad (4.9)$$

$$Prob(Z_t = k) = \sum_{i=1}^{m}\sum_{j=1}^{l} t_{i,j,k} \times Prob(X_t = i, Y_t = j)$$

where $Prob(Y_t = j, Z_t = k)$ is the joint probability distribution of Y_t and Z_t, $Prob(X_t = i, Z_t = k)$ is the joint probability distribution of X_t and Z_t, and $Prob(X_t = i, Y_t = j)$ is the joint probability distribution of X_t and Y_t.

To obtain x_i, y_j, and z_k, we assume that X_t, Y_t, and Z_t are all independent from each other which can be denoted as below:

$$Prob(X_t = i, Y_t = j) = Prob(X_t = i)Prob(Y_t = j)$$
$$Prob(X_t = i, Z_t = k) = Prob(X_t = i)Prob(Z_t = k) \qquad (4.10)$$
$$Prob(Y_t = j, Z_t = k) = Prob(Y_t = j)Prob(Z_t = k)$$

Consequently, through combining the equations with the assumptions above, we get:

$$x_i = \sum_{j=1}^{l}\sum_{k=1}^{n} f_{i,j,k} y_j z_k, \, i = 1, 2, \ldots, m,$$

$$y_j = \sum_{i=1}^{m}\sum_{k=1}^{n} r_{i,j,k} x_i z_k, \, j = 1, 2, \ldots, l, \qquad (4.11)$$

$$z_k = \sum_{i=1}^{m}\sum_{j=1}^{l} t_{i,j,k} x_i y_j, \, k = 1, 2, \ldots, n.$$

The equations above can be written in a tensor format:

$$x = Fyz, \, y = Rxz, \, z = Txy \qquad (4.12)$$

with $\sum_{i=1}^{m} x_i = 1$, $\sum_{j=1}^{l} y_j = 1$, and $\sum_{k=1}^{n} z_k = 1$.

According to the analysis above, we can design the following algorithm to co-rank the importance of objects and relations.

Algorithm 4.1 HRank-CO Algorithm

Input:
  Three tensors F, T and R, three initial probability distributions x_0, y_0 and z_0 and the tolerance ε.
Output:
  Three stationary probability distributions x, y and z.
Procedure:
Set $t = 1$;
repeat
    Compute $x_t = F y_{t-1} z_{t-1}$;
    Compute $y_t = R x_t z_{t-1}$;
    Compute $z_t = T x_t y_t$;
until $||x_t - x_{t-1}|| + ||y_t - y_{t-1}|| + ||z_t - z_{t-1}|| < \varepsilon$

4.1.3   Experiments

In this section, we do experiments to validate the effectiveness of three versions of HRank on three real datasets, respectively. Here we use three real datasets: DBLP dataset [14, 25], ACM dataset [14], and IMDB dataset [16].

4.1.3.1   Ranking of Homogeneous Objects

Since the homogeneous objects are connected by symmetric constrained meta paths, the experiments validate the effectiveness of HRank-SY on symmetric constrained meta paths.

Experiment Study on Symmetric Constrained Meta Paths This experiment ranks the same type of objects by designating a symmetric constrained meta path on ACM dataset. Here, we rank the importance of authors through the symmetric meta path *APA*, which considers the co-author relations among authors. We also employ two constrained meta paths *APA*|*P.L* = "*H*.2" and *APA*|*P.L* = "*H*.3", where the categories of ACM *H*.2 and *H*.3 represent "database management" and "information storage/retrieval," respectively. That is, two constrained meta paths subtly consider the co-author relations in database/data mining field and information retrieval field, respectively. We employ HRank-SY to rank the importance of authors based on these three paths. As the baseline methods, we rank the importance of authors with PageRank and the degree of authors (called Degree method). We directly run PageRank on the whole ACM network by ignoring the heterogeneity of objects. Since the results of PageRank mix all types of objects, we select the author type from the ranking list as the final results.

The top ten authors of each method are shown in Table 4.1. We can find that all these ranking lists have some common influential authors except that of PageRank. The results of PageRank include some not very well-known authors in database/information retrieval (DB/IR) field, such as Ming Li and Wei Wei, although they may be very influential in other fields. We know that the PageRank values of

Table 4.1 Top ten authors of different methods on ACM dataset. The number in the parenthesis of the fifth column means the rank of authors in the whole ranking list returned by PageRank

Rank	APA	APA|P.L = "H.3"	APA|P.L = "H.2"	PageRank	Degree
1	Jiawei Han	W. Bruce Croft	Jiawei Han	Ming Li(1522)	Jiawei Han
2	Philip Yu	ChengXiang Zhai	Christos Faloutsos	Wei Wei(2072)	Philip Yu
3	Christos Faloutsos	James Allan	Philip Yu	Jiawei Han(5385)	ChengXiang Zhai
4	Zheng Chen	Jamie Callan	Jian Pei	Tao Li(6090)	Zheng Chen
5	Wei-Ying Ma	Zheng Chen	H. Garcia-Molina	Hong-Jiang Zhang(6319)	Christos Faloutsos
6	ChengXiang Zhai	Ryen W. White	Jeffrey F. Naughton	Wei Ding(6354)	Ravi Kumar
7	W. Bruce Croft	Wei-Ying Ma	Divesh Srivastava	Jiangong Zhang(7285)	W. Bruce Croft
8	Scott Shenker	Jian-Yun Nie	Raghu Ramakrishnan	Christos Faloutsos(7895)	Wei-Ying Ma
9	H. Garcia-Molina	Gerhard Weikum	Charu C. Aggarwal	Feng Pan(8262)	Gerhard Weikum
10	Ravi Kumar	C. Lee Giles	Surajit Chaudhuri	Hongyan Liu(8440)	Divesh Srivastava

objects are decided by their degrees to a large extent, so the rank values of affiliation objects are high due to their high degrees. It improves the rank values of author objects connecting multiple high-ranking affiliations. The bad results of PageRank show that the ranking in heterogeneous networks should consider the heterogeneity of objects. Otherwise, it cannot distinguish the effect of different types of links. Moreover, we can also observe that the results of HRank with constrained meta paths have obvious bias on the field it assigns. For example, the path $APA|P.L = $ "H.3" reveals the important authors in information retrieval field, such as W. Bruce Croft, ChengXiang Zhai, and James Allan. However, the path $APA|P.L = $ "H.2" returns the influential authors in database and data mining field, such as Jiawei Han and Christos Faloutsos. For the meta path APA, it mingles well-known authors in these two fields. The results illustrate that the constrained meta paths are able to capture subtle semantics by deeply disclosing the most influential authors in a certain field.

Quantitative Comparison Experiments Based on the results returned by five methods, we can obtain five candidate ranking lists of authors in ACM dataset. To evaluate the results quantitatively, we crawled data as ground truth from two well-known websites. The first ground truth provides the author ranks from Microsoft Academic Search.[1] Specifically, we crawled two standard ranking lists of authors in two

---

[1] http://academic.research.microsoft.com/.

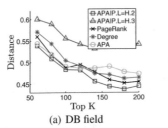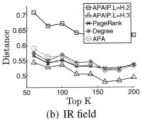

(a) DB field   (b) IR field

Fig. 4.4 The distances between the ranking lists obtained by different methods and the standard ranking lists on different fields on ACM dataset. The ground truth is from Microsoft Academic Search

academic fields: DB and IR. Then, we compare the difference between our candidate ranking lists and the standard ranking lists. In order to measure the quality of the ranking results, we use the *Distance* criterion proposed in [12], which is defined as follows.

$$D(R, R') = \frac{\sum_{i=1}^{n} [(n-i) \times \sum_{j=1 \wedge R'_j \notin \{R_1, \dots, R_i\}}^{i} 1]}{\sum_{i=1}^{\lfloor \frac{n}{2} \rfloor} [(n-i) \times i] + \sum_{i=\lfloor \frac{n}{2} \rfloor+1}^{n} [(n-i) \times (n-i)]} \tag{4.13}$$

where R_i represents the ith object in ranking list R, while R'_j denotes the jth object in ranking list R'. And n is the total number of objects in the ranking lists. Note that the numerator of the formula measures the real distance between the two rankings, and the denominator of the formula is used to normalize the real distance to a number between 0 and 1. So the criterion not only measures the number of mismatches between these two lists, but also considers the position of these mismatches. The smaller *Distance* means the smaller difference (i.e., better performance).

In this experiment, we compare the five candidate ranking lists with each of the two standard ranking lists from Microsoft Academic Search and the *Distance* results are shown in Fig. 4.4. We can observe an obvious phenomenon: The results obtained by the constrained meta paths have the smallest *Distance* on its corresponding field, while they have the largest *Distance* on other fields. For example, HRank with the path $APA|P.L = \text{"}H.2\text{"}$ has the smallest *Distance* on the DB field in Fig. 4.4a, while it has the largest *Distance* on IR field in Fig. 4.4b. The reason lies in that the path $APA|P.L = \text{"}H.2\text{"}$ focuses on the authors in the DB field. Meanwhile, these authors deviate from those in the IR field. The results further illustrate that the constrained meta path can disclose the influential authors in a certain field more correctly. Since the meta path (i.e., *APA*) considers the co-author relationship on all fields, it achieves mediocre performances on these two fields. In fact, the HRank with meta path *APA* only achieves closer performances to PageRank and Degree methods. It implies that the constrained meta path in HRank indeed helps to improve the ranking performances in a specific field.

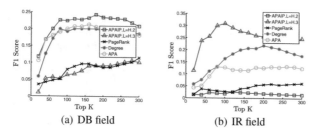

Fig. 4.5 F1 accuracy of the ranking lists obtained by different methods on different fields on ACM dataset. The ground truth is from ArnetMiner

Furthermore, we quantitatively evaluate the results according to the second ground truth from ArnetMiner [26] that offers comprehensive search and mining services for academic community.[2] Specifically, we crawl the first 200 authors as experts in DB and IR fields through searching "data mining" and "information retrieval." Since these 200 experts have no ranking order, we evaluate the accuracy of the top k authors of five candidate ranking lists with the F1 score. From the results shown in Fig. 4.5, we can observe the same phenomenon. That is, the constrained meta paths always achieve the best performances on their corresponding fields, while they have the worst performances on other fields (note that the higher F1 score means the better performances). Moreover, the meta paths also have the moderate performances. The experiments on both ground truths confirm that HRank is able to improve the ranking performances in a specific field through assigning constrained meta paths.

4.1.3.2 Ranking of Heterogeneous Objects

Then, the experiments validate the effectiveness of HRank-AS on asymmetric constrained meta paths.

Experiment Study on Asymmetric Constrained Meta Paths The experiments are done on the DBLP dataset. We evaluate the importance of authors and conferences simultaneously based on the meta path APC, which means authors publish papers on conferences. Two constrained meta paths ($APC|P.L = $ "DB" and $APC|P.L = $ "IR") are also included, which means authors publish DB(IR) field papers on conferences. Similarly, the experiments also include two baseline methods (i.e., PageRank and Degree) in above experiments with the same experimental process.

The top ten authors and conferences returned by these five methods are shown in Tables 4.2 and 4.3, respectively. As shown in Table 4.2, the ranking results of these methods on authors all are reasonable; however, the constrained meta paths can find the most influential authors in a certain field. For example, the top three authors of $APC|P.L = $ "DB" are Surajit Chaudhuri, Hector Garcia-Molina, and H.V. Jagadish, and all of them are very influential researchers in the database field. The

---

[2]http://arnetminer.org/.

Table 4.2 Top ten authors of different methods on DBLP dataset. The number in the parenthesis of the fifth column means the rank of authors in the whole ranking list returned by PageRank

Rank	APC	APC|P.L = "DB"	APC|P.L = "IR"	PageRank	Degree
1	Gerhard Weikum	Surajit Chaudhuri	W. Bruce Croft	W. Bruce Croft(23)	Philip S. Yu
2	Katsumi Tanaka	H. Garcia-Molina	Bert R. Boyce	Gerhard Weikum(24)	Gerhard Weikum
3	Philip S. Yu	H.V. Jagadish	Carol L. Barry	Philip S. Yu(25)	Divesh Srivastava
4	H. Garcia-Molina	Jeffrey F. Naughton	James Allan	Jiawei Han(26)	Jiawei Han
5	W. Bruce Croft	Michael Stonebraker	ChengXiang Zhai	H. Garcia-Molina(27)	H. Garcia-Molina
6	Jiawei Han	Divesh Srivastava	Mark Sanderson	Divesh Srivastava(28)	W. Bruce Croft
7	Divesh Srivastava	Gerhard Weikum	Maarten de Rijke	Surajit Chaudhuri(29)	Surajit Chaudhuri
8	Hans-Peter Kriegel	Jiawei Han	Katsumi Tanaka	H.V. Jagadish(30)	H.V. Jagadish
9	Divyakant Agrawal	Christos Faloutsos	Iadh Ounis	Jeffrey F. Naughton(31)	Jeffrey F. Naughton
10	Jeffrey Xu Yu	Philip S. Yu	Joemon M. Jose	Rakesh Agrawal(32)	Rakesh Agrawal

Table 4.3 Top ten conferences of different methods on DBLP dataset. The number in the parenthesis of the fifth column means the rank of conferences in the whole ranking list returned by PageRank

Rank	APC	APC|P.L = "DB"	APC|P.L = "IR"	PageRank	Degree
1	CIKM	ICDE	SIGIR	ICDE(3)	ICDE
2	ICDE	VLDB	WWW	SIGIR(4)	SIGIR
3	WWW	SIGMOD	CIKM	VLDB(5)	VLDB
4	VLDB	PODS	JASIST	CIKM(6)	SIGMOD
5	SIGMOD	DASFAA	WISE	SIGMOD(7)	CIKM
6	SIGIR	EDBT	ECIR	JASIST(8)	JASIST
7	DASFAA	ICDT	APWeb	WWW(9)	WWW
8	JASIST	MDM	WSDM	DASFAA(10)	PODS
9	WISE	WebDB	JCIS	PODS(11)	DASFAA
10	EDBT	SSTD	IJKM	JCIS(12)	EDBT

top three authors of $APC|P.L = "IR"$ are W. Bruce Croft, Bert R. Boyce, and Carol L. Barry, and they all have the high academic reputation in the information retrieval field. Similarly, as we can see in Table 4.3, HRank with constrained meta paths (i.e., $APC|P.L = "DB"$ and $APC|P.L = "IR"$) can clearly find the important conferences in DB and IR fields, while other methods mingle these conferences. For example, the most important conferences in the DB field are ICDE, VLDB, and SIGMOD, while the most important conferences in the IR field are SIGIR, WWW, and CIKM. Observing Tables 4.2 and 4.3, we can also find the mutual effect of authors and conferences. That is, an influential author published many papers in the important conferences, and vice versa. For example, W. Bruce Croft published many papers in SIGIR and CIKM, while Surajit Chaudhuri has many papers in SIGMOD, ICDE, and VLDB.

Quantitative Comparison Experiments To verify the effectiveness of these methods, we use the above *Distance* criterion to calculate the difference between their results and standard ranking lists crawled from Microsoft Academic Search. Figure 4.6 shows the differences of author ranking lists. We can observe the same phenomenon with above quantitative experiments again. That is, HRank with constrained meta paths achieve the best performances on their corresponding field. Meanwhile, they have the worst performances on other fields. In addition, compared to that of PageRank and Degree, the mediocre performances of HRank with meta path APC further demonstrate the importance of constrained meta path to capture the subtle semantics contained in heterogeneous networks. Similarly, we further evaluate the F1 accuracy of these methods according to the ground truth crawled from ArnetMiner. The results are shown in Fig. 4.7. Once again the results reveal the same findings that HRank can more accurately discover the authors ranking in a special field with the help of constrained meta path.

Experiments on Meta Path with Multiple Constraints Furthermore, we validate the effectiveness of meta path with multiple constraints. In the above experiments, we employ the constraint on the label of papers in HRank with the meta path APC. Here, we add one more constraint on conference. Specifically, by contrast to the constrained meta path $APC|P.L = "DB"$, we employ the paths $APC|P.L =$

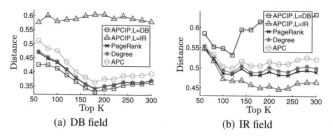

(a) DB field   (b) IR field

Fig. 4.6 Distances between the candidate author ranking lists and the standard ranking lists on different fields on DBLP dataset. The ground truth is from Microsoft Academic Search

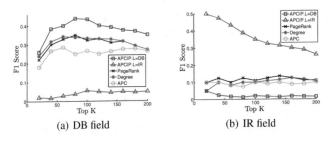

Fig. 4.7 F1 accuracy of the ranking lists obtained by different methods on different fields on DBLP dataset. The ground truth is from ArnetMiner

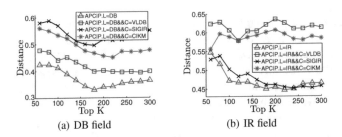

Fig. 4.8 Rank accuracy of HRank with different constrained meta paths on DBLP dataset

"*DB*"&&*C* = "*VLDB*", *APC*|*P.L* = "*DB*"&&*C* = "*SIGIR*", and *APC*|*P.L* = "*DB*"&&*C* = "*CIKM*", which mean authors publish DB field papers on specified conferences (e.g., VLDB, SIGIR, and CIKM). Similarly, we add the same conference constraints on the path *APC*|*P.L* = "*IR*". Same with the above experiments, we calculate the rank accuracy of HRank with these constrained meta paths and the results are shown in Fig. 4.8.

We know that HRank with the path *APC*|*P.L* = "*DB*" (*APC*|*P.L* = "*IR*") can reveal the influence of authors in the DB (IR) field. As ground truth, this ranking is based on the aggregation of many conferences related to the DB field. The added conference constraint in HRank further reveals the influence of authors in the specific conference of the field. So we can use the closeness to the ground truth to reveal the importance of a conference to that field. That is, if the ranking from a specific conference is quite closer to the ground truth rank, that can imply the conference is a dominating conference in that field. From Fig. 4.8a, we can find that the VLDB conference constraint (the blue curve) achieves the closest performances to the ground truth ranking, while the performances of the SIGIR conference constraint (the black curve) deviate most. So we can infer that the VLDB is more important than SIGIR in the DB field and the CIKM has the middle importance. Similarly, from Fig. 4.8b, we can infer that the SIGIR is more important than VLDB in the IR field. These findings comply with our common sense. As we know, although the VLDB and SIGIR both are the top conferences in computer science, they are very important only in their research fields. For example, the VLDB is important in the DB field, while it is not so important in the IR field. The middle importance of the CIKM conference stems from the fact that it is a comprehensive conference including papers from both DB and IR fields. In addition, we can find that the SIGIR curve almost overlaps with the ground truth over the IR field, while the VLDB curve still has a gap with the ground truth over the DB field. We think the reason is that SIGIR is the main conference in the IR field, while in the DB field, there are also other important conferences, such as SIGMOD and ICDE. Overall, the experiments show that HRank with constrained meta path can not only effectively find the influential authors in each research field on a specified conference but also indirectly reveal the importance of conferences in the fields. It also implies that HRank can achieve accurate and subtle ranking results by flexibly setting the combination of constraints.

4.1.3.3 Co-ranking of Objects and Paths

Experiment Study on Co-ranking on Symmetric Constrained Meta Paths In this experiment, we will validate the effectiveness of HRank-CO to rank objects and symmetric constrained meta paths simultaneously. The experiment is done on ACM dataset. First, we construct a $(2, 1)$th order tensor X based on 73 constrained meta paths (i.e., $APA|P.L = L_j, j = 1 \cdots 73$). When the ith and the kth authors co-publish a paper together, of which the label is the jth label (i.e., ACM categories), we add one to the entries $x_{i,j,k}$ and $x_{k,j,i}$ of X. In this case, X is symmetric with respect to the index j. By considering all the publications, $x_{i,j,k}$ (or $x_{k,j,i}$) refers to the number of collaborations by the ith and the kth author under the jth paper label. In addition, we do not consider any self-collaboration, i.e., $x_{i,j,i} = 0$ for all $1 \leq i \leq 17,431$ and $1 \leq j \leq 73$. The size of X is $17,431 \times 73 \times 17,431$ where there are 91,520 nonzero entries in X. The percentage of nonzero entries is $4.126 \times 10^{-4}\%$. In this dataset, we will evaluate the importance of authors through the co-author relations, and meanwhile, we will analyze the importance of paths (i.e., which paths have the most contributions to the importance of authors).

Figure 4.9 shows the stationary probability distributions of authors and paths. It is obvious that some authors and paths have higher stationary probability, which implies

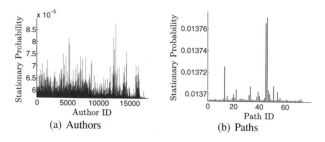

(a) Authors   (b) Paths

Fig. 4.9 Stationary probability distributions of authors and constrained meta paths

Table 4.4 Top 10 authors and constrained meta paths (note that only the constraint (L_j) of the paths $(APA|P.L = L_j, j = 1 \ldots 73)$ are shown in the third column of the table)

Rank	Authors	Constrained meta paths
1	Jiawei Han	H.3 (Information storage and retrieval)
2	Philip Yu	H.2 (Database management)
3	Christos Faloutsos	C.2 (Computer-communication networks)
4	Ravi Kumar	I.2 (Artificial intelligence)
5	Wei-Ying Ma	F.2 (Analysis of algorithms and problem complexity)
6	Zheng Chen	D.4 (Operating systems)
7	Hector Garcia-Molina	H.4 (Information systems applications)
8	Hans-Peter Kriegel	G.2 (Discrete mathematics)
9	Gerhard Weikum	I.5 (Pattern recognition)
10	D.R. Karger	H.5 (Information interfaces and presentation)

that these authors and paths are more important than others. Table 4.4 shows the top ten authors (left) and paths (right) based on their HRank values. We can find that the top ten authors all are influential researchers in the DM/IR fields, which conforms to our common senses. Similarly, the most important paths are related to DM/IR fields, such as $APA|P.L = $ "H.3" (information storage and retrieval) and $APA|P.L = $ "H.2" (database management). Although the conferences in ACM dataset are from multiple fields, such as DM/DB (e.g., KDD, SIGMOD) and computation theory (e.g., SODA, STOC), there are more papers from the DM/DB fields, which makes the authors and paths in the DM/DB fields ranked higher. We can also find that the influence of authors and paths can be promoted by each other. The reputation of Jiawei Han and Philip Yu come from their productive papers in the influential fields (e.g., H.3 and H.2). In order to observe this point more clearly, we show the number of co-authors of the top ten authors based on the top ten paths in Table 4.5. We can observe that there are more collaborations for top authors in the influential fields. For example, although Zheng Chen (rank 6) has more number of co-authors than Jiawei Han (rank 1), the collaborations of Jiawei Han focus on ranked higher fields (i.e., H.3 and H.2), so Jiawei Han has higher HRank score. Similarly, the top paths contain many collaborations of influential authors.

Experiment Study on Co-ranking on Asymmetric Constrained Meta Paths The experiments on the Movie dataset aim to show the effectiveness of HRank-CO to rank heterogeneous objects and asymmetric constrained meta paths simultaneously. In this case, we construct a third-order tensor X based on the constrained meta paths $AMD|M.T$. That is, the tensor represents the actor–director collaboration relations on different types of movies. When the ith actor and the kth director cooperate in a movie of the jth type, we add one to the entries $x_{i,j,k}$ of X. By considering all the cooperations, $x_{i,j,k}$ refers to the number of collaborations by the ith actor and the kth director under the jth type of movie. The size of X is $5324 \times 112 \times 551$, and there are 36,529 nonzero entries in X. The percentage of nonzero entries is $7.827 \times 10^{-4}\%$.

Table 4.5 Number that the top ten authors collaborate with others via the top ten constrained meta paths (note that only the constraints (L_j) of the paths ($APA|P.L = L_j, j = 1 \ldots 73$) are shown in the first row of the table)

Ranked author/CP	1 (H.3)	2 (H.2)	3 (C.2)	4 (I.2)	5 (F.2)	6 (D.4)	7 (H.4)	8 (G.2)	9 (I.5)	10 (H.5)
1 (Jiawei Han)	51	176	0	0	0	0	9	2	2	0
2 (Philip Yu)	51	94	0	0	9	0	3	0	13	0
3 (C. Faloutsos)	17	107	0	5	9	0	3	4	2	0
4 (Ravi Kumar)	73	27	0	3	13	0	18	5	0	0
5 (Wei-Ying Ma)	132	26	0	9	0	0	2	0	30	10
6 (Zheng Chen)	172	9	0	9	0	0	22	0	38	9
7 (H. Garcia-Molina)	23	65	3	0	0	0	1	0	0	4
8 (H. Kriegel)	19	28	5	0	0	0	6	0	7	4
9 (G. Weikum)	82	14	0	4	0	0	8	0	4	0
10 (D.R. Karger)	11	5	13	0	7	4	1	7	0	7

Table 4.6 Top 10 actors, directors, and meta paths on IMDB dataset (note that only the constraints (T_j) of the paths ($AMD|M.T = T_j, j = 1 \ldots 1591$) are shown in the fourth column)

Rank	Actor	Director	Conditional meta path
1	Eddie Murphy	Tim Burton	Comedy
2	Harrison Ford	Zack Snyder	Drama
3	Bruce Willis	Marc Forster	Thriller
4	Drew Barrymore	David Fincher	Action
5	Nicole Kidman	Michael Bay	Adventure
6	Nicolas Cage	Ridley Scott	Romance
7	Hugh Jackman	Richard Donner	Crime
8	Robert De Niro	Steven Spielberg	Sci-Fi
9	Brad Pitt	Robert Zemeckis	Animation
10	Christopher Walken	Stephen Sommers	Fantasy

Table 4.6 shows the top ten actors, directors, and constrained meta paths (i.e., movie type). We observe the mutual enhancements of the importance of objects and meta paths again. Basically, the results comply with our common senses. The top ten actors are well known, such as Eddie Murphy and Harrison Ford. Similarly, these directors are also famous in filmdom due to their works. These movie types obtained are the most popular movie subjects as well. In addition, we can observe the mutual effect of objects and paths one more time. As we know, Eddie Murphy and Drew Barrymore (rank 1, 4 in actors) are famous comedy and drama (rank 1, 2 in paths) actors. Harrison Ford and Bruce Willis (rank 2, 3 in actors) are popular thrill and action (rank 3, 4 in paths) actors. These higher ranked directors also prefer to those popular movie subjects. Furthermore, we also compare these results with the

recommended results from the IMDB website.[3] Although only a subset of movies in IMDB is included in our experiments, the 80% of the top 10 actors in our results are included in the set of the top 250 greatest movie actors in all time recommended by IMDB,[4] and the 50% of the top 10 directors in our results are included in the set of the top 50 favorite directors recommended by IMDB.[5] Moreover, most of movie types recommended by our method have high ranks in the popular types summarized by IMDB.[6] The more details of the HRank method and experimental results can be seen in [18].

4.2 Ranking-Based Clustering

4.2.1 Overview

Recently, the link-based clustering attracts more and more attention, which usually groups objects that are densely interconnected but sparely connected with the rest of the network [11]. Also with the boom of search engine, object ranking [1, 5] becomes an important data mining task, which evaluates the importance of objects. Conventionally, clustering and ranking are two independent tasks and they are usually used separately. However, recent researches show that clustering and ranking can mutually promote each other and their combination makes more sense in many applications [21, 22]. If we know the important objects in a cluster, we can understand this cluster better; and the ranking in a cluster provides more subtle and meaningful information for clustering. Although it is a promising way to do clustering and ranking together, previous approaches are confined to a simple HIN with special structure. For example, Sun et al. validated the mutual improvement of clustering and ranking in bipartite network [21] (an example shown in Fig. 4.10a) and star-schema network [22] (an example shown in Fig. 4.10b). Shi et al. [27] integrated clustering and ranking in the hybrid network including heterogeneous and homogeneous relations. However, the data in real applications are usually more complex and irregular, which are beyond the widely used bipartite or star-schema network. For example, the bibliographic data (see an example in Fig. 4.10c) include not only heterogeneous relations but also homogeneous relations (e.g., self loop on P); the bioinformatics data [2] (see an example in Fig. 4.10d) have more complex structure, which includes multiple hub objects (e.g., C and G). So it is desirable to design effective ranking-based clustering algorithm for these complex and irregular HIN data. Broadly speaking, for HIN with arbitrary schema, we need to design a general solution to manage the objects and their relations, which is the basic for mining useful patterns on it.

---

[3]http://www.imdb.com/.

[4]http://www.imdb.com/list/ls050720698/.

[5]http://www.imdb.com/list/ls050131440/.

[6]http://www.imdb.com/list/ls050782187/?view=detail&sort=listorian:asc.

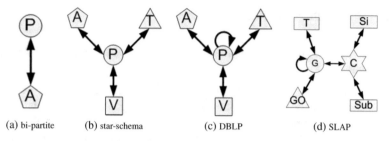

(a) bi-partite (b) star-schema (c) DBLP (d) SLAP

Fig. 4.10 Examples of heterogeneous information networks. The letters are the abbreviation of different types of objects (e.g., P: paper, A: author)

Obviously, it is more practical and useful to determine the underlying clusters and ranks on a general heterogeneous information network, but they are seldom exploited until now. When we integrate ranking and clustering on an HIN with arbitrary schema, it faces the following challenges. (1) A general HIN has more complex structure. For a simple HIN with a bipartite or star-schema structure, it is relatively easy to manage heterogeneous objects and build models. However, a general HIN may have arbitrary schema, beyond the bipartite or star-schema structure. Although an intuitive way is to decompose it into multiple simpler subnetworks, the issue is how we decompose the HIN without structural information loss and maintain the consistency among the decomposed subnetworks. (2) It is challenging to integrate the clustering and ranking in a complex heterogeneous network. We know that it is still a daunting task to separately do clustering and ranking on a general HIN. Therefore, it is more difficult to design an effective mechanism to combine these two tasks on the HIN.

In this chapter, we study the ranking-based clustering problem on a general HIN and propose a novel algorithm **HeProjI** to solve the **He**terogeneous network **Proj**ection and **I**ntegration of clustering and ranking tasks. In order to conveniently manage objects and relations in an HIN with arbitrary schema, we design a network projection method to project the HIN into a sequence of subnetworks without structural information loss, where the subnetwork may be a relatively simple bipartite or star-schema network. Moreover, an information transfer mechanism is developed to maintain the consistency across subnetworks. For each subnetwork, a path-based random walk method is proposed to generate the reachable probability of objects, which can be effectively used to estimate the cluster membership probability and the importance of objects. Through iteratively analyzing each subnetwork, HeProjI can obtain the steady and consistent clustering and ranking results. We perform a number of experiments on three real datasets to validate the effectiveness of HeProjI. The results show that HeProjI not only achieves better clustering and ranking accuracy compared to well-established algorithms, but also effectively handles complex HIN which cannot be handled by previous methods.

4.2.2 Problem Formulation

In this section, we give the problem definition and some important concepts used in this chapter.

Definition 4.7 (*General heterogeneous information network*) Given a schema A = (T, R) which consists of a set of entities type T = {T} and a set of relations R = {R}, a general information network is defined as a graph G = (X,E) with an object type mapping function $\tau : X \rightarrow$ T and link type mapping function $\psi : E \rightarrow$ R. Each object |T| > 1 or the types of relations |R| > 1, the network is called **heterogeneous information network**; otherwise, it is a **homogeneous information network**.

Figure 4.10 shows the schema of several HIN examples. The bipartite network in Fig. 4.10a only includes two types of objects, and the widely used star-schema network [16, 22, 24] in Fig. 4.10b organizes objects in HIN with one target type and several attribute types. However, a general heterogeneous information network may be more complex and irregular. It may not only include homogeneous or heterogeneous relations, but also include multiple hub objects. Figure 4.10d shows such a general HIN example. The object G has heterogeneous relations (e.g., $G{\rightarrow}GO$ and $G{\rightarrow}C$) as well as homogeneous relations (e.g., $G{\rightarrow}G$). Moreover, the network is beyond the star-schema because of multiple hub objects (e.g., G and C). It is clear that bipartite graph and star-schema network are the special case of a general HIN.

For a general HIN, it is difficult to manage objects and relations in the network. Although we can project it into several homogeneous networks through assigning meta paths as reference [3] did, it will loss much information among different-typed objects. We know that, as the special case of HIN, the bipartite and star-schema networks are relatively easy to manage objects and relations in the network. So a basic idea of handling a general HIN is to decompose it into simpler networks. Following this idea, we design a novel HIN projection method. Specifically, we can select one type (called pivotal type) and its connected other types (called supportive type). These types and their relations constitute the schema of a projected subnetwork of original HIN. Formally, it can be defined as follows:

Definition 4.8 (*Projected subnetwork*) For an HIN with schema A = (T, R), its projected subnetwork has the schema $A' = (T', R')$ where $T' \subset$ T, $R' \subset$ R, T' includes one **pivotal type** (denoted as P) and other types connected with P (called **supportive type**, denoted as S = {S}). R' includes the heterogeneous relations between P and S and homogeneous relations among P (if existing).

A projected subnetwork can be denoted as $P - S$. The $X^{(P)}$ is the object set of pivotal type, and $X^{(S)}$ represents the object set of supportive type S. For convenience, the projected subnetwork is also called subnetwork which can be represented with its pivotal type P. For example, Fig. 4.11c shows the projected subnetwork $G - \{C, T, GO\}$ with type G object (the one in red) as the pivotal type, while types C, T and GO are the supportive types as they are object types connected to object type G. Similarly, Fig. 4.11b and d shows the projected subnetworks with pivotal type objects GO and C, respectively.

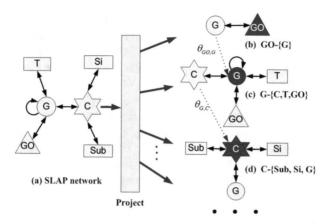

Fig. 4.11 An example of HIN projection. The pivotal type is marked with *red color*. The *dot line* represents the information transfer among subnetworks

It is clear that an HIN can be projected into a sequence of subnetworks through selecting different pivotal types. So we define the HIN projection concept as follows.

Definition 4.9 (*HIN projection*) An HIN with t types of objects can be projected into an ordered set of t projected subnetworks by successively selecting one of the t types as pivotal type.

Figure 4.11 shows a projection example of SLAP network, a bioinformatics dataset [17]. Through successively selecting the six object types (*GO, G, C* and so on) as pivotal type, the SLAP network is projected into a sequence of six subnetworks. It is clear that the HIN projection has the following properties.

Property 4.1 *HIN projection is a structure–information-lossless network decomposition.*

According to Definition 4.9, all objects and relations in original HIN are in the projected subnetworks. That is to say, the HIN can be reconstructed from the set of projected subnetworks.

Property 4.2 *Each projected subnetwork in HIN projection should be a bipartite graph or a star-schema network (with self loop).*

According to Definition 4.8, if there are two types of objects in the subnetwork, it is a bipartite graph; otherwise, it is a star-schema network. Note that, different from the conventional bipartite and star-schema networks, the pivotal type in subnetworks may include the homogenous relation (i.e., self loop).

Property 4.3 *HIN projection is not unique for a general HIN.*

An HIN has different projection sequences through selecting different orders of pivotal types. For example, the SLAP network in Fig. 4.11 has the projection sequences: $GO - G - C - Si - Sub - T$, $T - G - GO - C - Si - Sub$, and so on. In fact, an HIN with t types of objects has the $t!$ projection sequences in all.

Assume that J represents a type in type set $\{T\}$. The object set can be denoted as $X=\{X^{(J)}\}$, and $X^{(J)}=\{X_p^{(J)}\}$ where $X_p^{(J)}$ is the object $p \in X^{(J)}$ (i.e., $\tau(p)=J$). The relations among objects include two types (homogeneous and heterogeneous relations), which can be represented by the two types of matrices, **homogeneous** and **heterogeneous relation matrices**, respectively. If type J has homogeneous relation (e.g., the self loop on P in Fig. 4.10c), the homogenous relation matrices can be written as $H^{(J)}$, where $H_{pq}^{(J)}$ denotes the relation between $X_p^{(J)}$ and $X_q^{(J)}$. If two types (I and J) have heterogeneous relation (e.g., $P - A$ in Fig. 4.10c), the heterogeneous relation matrices can be written as $H^{(I,J)}$, where $H_{pq}^{(I,J)}$ denotes the relation between $X_p^{(I)}$ and $X_q^{(J)}$. Correspondingly, we have **homogeneous transition matrix $M^{(J)}$** and **heterogeneous transition matrix $M^{(I,J)}$**. It is clear that the transition matrix $M^{(I,J)}$ can be derived from the relation matrix $H^{(I,J)}$ by $M^{(I,J)}=D^{(I,J)^{-1}}H^{(I,J)}$, where $D^{(I,J)}$ is the diagonal matrix with the diagonal value equaling to the corresponding row sum of $H^{(I,J)}$. Similarly, $M^{(J)}=D^{(J)^{-1}}H^{(J)}$. Taking Fig. 4.10c as example, $M^{(P)}$ is the transition probability matrix of the citation relation $H^{(P)}$, and $M^{(A,P)}$ is the transition probability matrix of the $A - P$ relation $H^{(A,P)}$. For given network structure, we can derive the homogeneous and heterogeneous transition matrices. In the following section, we consider that the transition matrices are known.

Different from conventional clustering in homogeneous networks, cluster in HIN should include different types of objects, where these objects share the same semantic meaning. For example, in bibliographic data, a cluster about data mining area includes venues, authors, and papers in this field. For each type of objects $X^{(J)}$, we define the **membership matrix $B^{(J|C_k)} \in [0, 1]^{|X^{(J)}| \times |X^{(J)}|}$**, which is a diagonal matrix whose diagonal value represents the membership probability of $X_p^{(J)}$ belonging to the cluster C_k. Note that the sum of membership probability of $X_p^{(J)}$ in K clusters is 1 (i.e., $\sum_{k=1}^{K} B_{pp}^{(J|C_k)}=1$). Now, we can formulate the problem of clustering on a general HIN as follows: Given a heterogeneous network $G=(X, E)$ and the semantic cluster number K, our goal is to find a clusters set $\{C_k\}_{k=1}^K$, where C_k is defined as $C_k = \{\{B^{(J|C_k)}\}_{J \in \{T\}}\}$. In this way, it is a soft clustering. That is, an object p in $X^{(J)}$ can belong to several clusters, and it is in a cluster C_k with the probability $B_{pp}^{(J|C_k)}$. Moreover, a cluster C_k can contain all kinds of objects.

4.2.3 The HeProjI Algorithm

Through the HIN projection, it will become much easier to analyze the HIN through handling a set of simple projected subnetworks, since these subnetworks are bipartite or star-schema networks. However, it may result in a troublesome business: how to maintain the consistency among different subnetworks. To solve it, we design an information transfer mechanism which inherits a portion of information from other subnetworks to current one. In order to integrate the clustering and ranking in a uniform framework, a model is required to flexibly support these two tasks. Following this idea, we build a probabilistic model to estimate the probability of supportive and pivotal objects in each subnetwork. Moreover, the probability of objects can effectively infer the clustering information and represent the importance of objects.

4.2.3.1 Framework of HeProjI Algorithm

Specifically, we first project the original HIN into a sequence of subnetworks and then randomly assign the pivotal objects of the first subnetwork into K clusters (i.e., initialize $\{C_k\}_{k=1}^{K}$). For each subnetwork, a path-based random walk method is proposed to estimate the reachable probability of supportive objects in each cluster C_k, and then, a generative model is used to obtain the probability of pivotal objects. After that, an EM algorithm is employed to estimate the posterior probability of objects (i.e., the clustering information $\{C_k\}_{k=1}^{K}$). According to probability of objects, we can also calculate their ranking in each cluster. The above step is repeated until convergence. In the iterative process, the clustering and ranking can mutually promote each other until they reach a steady result. The basic framework of HeProjI is shown in Algorithm 4.2. In the following sections, we will present these operations in detail.

4.2.3.2 Reachable Probability Estimation of Objects

Basic idea As we have noted that the built probabilistic model can not only support the clustering and ranking tasks but also maintain the consistency among subnetworks, so the design of the model should obey the following two rules: (1) PageRank principle. In order to support the ranking task, the probability of objects should be able to

---

Algorithm 4.2 HeProjI: Detecting K clusters on HIN

Input:
Cluster number K and transition probability matrix M.
Output:
Membership probability $B^{(J|C_k)}$ of objects on each cluster $\{C_k\}_{k=1}^{K}$

Project the HIN into a sequence of sub-networks
Randomly initialize the membership probability $B^{(J|C_k)}$
repeat
  Select the projected sub-network $(P - S)$ in order
  for cluster $C_k \in C$ **do**
    Establish the probability of supportive objects: $Pr(X^{(S)}|C_k)$
    Generate the probability of pivotal objects: $P(X^{(P)}|C_k)$
    Estimate the posterior probability of objects: $P(C_k|X^{(P)}), P(C_k|X^{(S)})$
  end for
  Rank the objects: $Rank(X^{(P)}|C_k), Rank(X^{(S)}|C_k)$
until the membership probability obtains convergence

---

reflect their ranks. In other words, the probability of objects should be positively correlated with the node degree. (2) Consistency principle. In order to maintain the consistency among subnetworks, an effective mechanism should be designed to transfer appropriate information among subnetworks.

For the first rule (i.e., PageRank principle), the random walk is an apparent solution. However, it is traditionally used in homogeneous networks [1, 5]. Although it is also used in bipartite graph [28], it is seldom applied in HIN. Sun et al. [22] employed it to estimate the probability of attribute objects in a star-schema network, while it is confined to two types of objects. Heterogeneous objects and link semantics make it difficult to directly employ random walk in HIN. In a projected subnetwork, there are different types of supportive objects and they are connected through pivotal objects. So the random walk among objects should follow the specified paths. That is, the random walkers among supportive objects would need to pass through the pivotal objects. As a consequence, we need to estimate the probability of supportive and pivotal objects separately. The reachable probability of a supportive object can be calculated as the sum of the probability of walkers from other supportive objects walking to it through the pivotal type. The probability of pivotal objects can be generated through its reachable supportive objects. Because the bipartite network only contains one supportive type, the probability of supportive object can be calculated by the sum of probability of walkers from the same type of objects walking to it through the pivotal type. Figure 4.12 shows the probability estimation process. The reachable probability of type C can be calculated by random walkers wandering from type GO and T to type C through type G in Fig. 4.12a.

For the second rule (consistency principle), it is an intuitive idea to transfer information among subnetworks. However, what and how do we transfer? It is clear that the subnetworks are overlapped. If we transfer the information of any overlapping types, the model may be hard to control, since two subnetworks may have many overlapping types and one type may appear in many subnetworks. If we do clustering on each subnetwork individually, it is difficult to map clusters among subnetworks. We know that the random walk among all supportive objects passes through the pivotal objects. So we only need to transfer the information of pivotal type, and then,

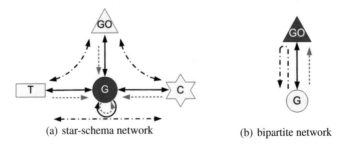

(a) star-schema network   (b) bipartite network

Fig. 4.12 Illustration of the probability estimation process for supportive and pivotal objects. The *black dash-dot line* represents the random walk process among supportive objects, and the *blue dotted line* represents the generative process of pivotal objects

the information can be propagated to other supportive objects by random walkers. In order to maintain the clustering consistency during the iteration, we let the pivotal objects in the current subnetwork inherit a portion of clustering information from previous subnetworks with a controlling parameter. The dot line in Fig. 4.11 shows two information inheritance examples. Specifically, the information on object G calculated in Fig. 4.11b is passed on to the calculation of the pivotal object G in Fig. 4.11c which affects the calculation of object C, while the information on object C is then passed on to the calculation of pivotal object C in Fig. 4.11d.

Reachable Probability for Supportive Objects First, we estimate the probability of supportive objects. The path-based random walk process is formulated with matrix representation. We use $M^{(S^I,S^J|P,C)}$ to represent the probability transition matrix from supportive type S^I to type S^J passing pivotal type P in the subnetwork C. $M^{(S^I,S^J|P,C)}$ can be calculated as follows:

$$M^{(S^I,S^J|P,C)} = M^{(S^I,P|C)} \times M^{(P,S^J|C)} \tag{4.14}$$

where $M^{(S^I,P|C)}$ is the transition matrix from S^I to P (i.e., $M^{(S^I,P)}$). Compared to conditional transition matrix $M^{(S^I,S^J|P,C_k)}$ defined below, $M^{(S^I,S^J|P,C)}$ is also called the global transition matrix, which is fixed for the subnetwork C. For example, in Fig. 4.12a, the global transition matrix $M^{(T,GO|G,C)}$ means the transition probability from type T to GO through G on the subnetwork $G - \{T, C, GO\}$. In the proposed model, the global probability of objects is important information to smooth the probability of pivotal objects (see Eq. 4.21 for more details).

When considering the clustering information, the transition matrices among supportive objects should be adjusted according to clusters. The clustering information can be represented by the membership matrix of pivotal objects, so the conditional transition matrix from S^I to S^J through P in the cluster C_k (i.e., $M^{(S^I,S^J|P,C_k)}$) can be defined as follows:

$$M^{(S^I,S^J|P,C_k)} = M^{(S^I,P|C)} \times B^{(P|C_k)} \times M^{(P,S^J|C)} \tag{4.15}$$

where $B^{(P|C_k)}$ is the membership of pivotal objects on cluster C_k.

The above transition matrices only consider the clustering information in the current subnetwork, which may cause the inconsistency among different subnetworks. For example, in the bibliographic data shown in Fig. 4.10c, clustering on the subnetwork $P - \{A, V, T\}$ may focus on research areas, while clustering on the subnetwork $A - \{P\}$ may more concern about co-author relations. In order to keep the clustering consistency among subnetworks, we can inherit a portion of cluster information from previous subnetworks. Only the clustering information of pivotal type is inherited from previous networks, and it is integrated with current clustering information of pivotal type. The reason why the simple mechanism work is that the pivotal objects, as hub node, can propagate the clustering information to all supportive objects. The transition matrices can be redefined as:

$$B''^{(P|C_k)} = \theta_{S,P} \times B'^{(P|C_k)} + (1 - \theta_{S,P}) \times B^{(P|C_k)} \tag{4.16}$$

$$M^{(S^I,S^J|P,C_k)} = M^{(S^I,P|C)} \times B''^{(P|C_k)} \times M^{(P,S^J|C)} \tag{4.17}$$

where $B'^{(P|C_k)}$ is the inherited membership matrix when the type P serves as a supportive type in the subnetwork whose pivotal type is S, and the $\theta_{S,P}$ is a learning rate parameter that controls the ratio of information inheritance from previous subnetwork (pivotal type is S) to current one (pivotal type is P). The dot line in Fig. 4.11 illustrates the two examples of information inheritance. The new transition matrix has the following advantages: (1) It transfers the clustering information among subnetworks, which keeps the consistency of subnetworks, and (2) it helps to speed up the convergence, since the priori clustering information is adopted. For a bipartite network, the transition probability matrix can be denoted as $M^{(S^I,S^J|P,C_k)}$, which has the same calculation mechanism.

The conditional probability of supportive type S^J on subnetwork C and cluster C_k is denoted as $Pr(X^{(S^J)}|C) \in [0,1]^{1 \times |X^{(S^J)}|}$ and $Pr(X^{(S^J)}|C_k) \in [0,1]^{1 \times |X^{(S^J)}|}$. Inspired by the PageRank [1], the probability of one type of objects is decided by the reachable probability from other types of objects through pivotal objects. So the conditional probability of supportive type S^J can be defined as follows.

$$Pr(X^{(S^J)}|C) = \sum_{S^I \in S, S^I \neq S^J} Pr(X^{(S^I)}|C) \times M^{(S^I,S^J|P,C)} \tag{4.18}$$

$$Pr(X^{(S^J)}|C_k) = \sum_{S^I \in S, S^I \neq S^J} Pr(X^{(S^I)}|C_k) \times M^{(S^I,S^J|P,C_k)} \tag{4.19}$$

The calculation is an iterative process, and $Pr(X^{(S^J)}|C_k)$ is initialized as the even value at the first iteration. For a bipartite network, random walkers start from type S^J and end up with the same type through the pivotal type P. The probability of supportive type S^J, $Pr(X^{(S^J)}|C_k)$ can be defined as $Pr(X^{(S^J)}|C_k) = Pr(X^{(S^J)}|C_k) \times M^{(S^J,S^J|C_k)}$.

Reachable Probability for Pivotal Objects Then, we estimate the probability of pivotal objects. We can consider that the pivotal objects are generated by adjacent supportive objects, so a generative model can be adopted here. The probability of pivotal objects comes from two parts: heterogeneous and homogeneous relations (if the pivotal type has self loop). For heterogeneous relations, the heterogeneous probability of pivotal object p in the subnetwork C (i.e., $Pr(X_p^{(P)}|C)$) can be calculated as follows:

$$Pr(X_p^{(P)}|C) = \prod_{S^J \in S} \prod_{q \in N(p)} Pr(X_q^{(S^J)}|C) \tag{4.20}$$

where $N(p)$ is the set of neighbors of object p in the subnetwork. It means that the pivotal object p is generated by the different types of adjacent supportive objects.

Then, we consider the probability of pivotal object p in a cluster C_k (i.e., $Pr(X_p^{(P)}|C_k)$). Similarly, the probability is also generated from the adjacent supportive objects in the cluster C_k. In addition, we add the global probability of pivotal object $X_p^{(P)}$ to smooth the probability:

$$Pr(X_p^{(P)}|C_k) = \lambda \prod_{S^J \in S} \prod_{q \in N(p)} Pr(X_q^{(S^J)}|C_k) + (1 - \lambda)Pr(X_p^{(P)}|C) \tag{4.21}$$

where the smooth parameter λ represents the portion of global probability. The smooth operation is an important component due to following reasons: (1) It prevents pivotal objects from accumulating into minority clusters, which helps to improve the clustering accuracy, and (2) it makes the probability change in pivotal objects more steady, which can improve the stability of HeProjI. The experiments in Sect. 5.7 also validate the importance of smooth operation.

For homogeneous relations (i.e., the pivotal object has self loop), we can calculate the cluster-based homogeneous transition probability for pivotal type as follows:

$$M^{(P|C_k)} = M^{(P|C)} \times B^{(P|C_k)} \tag{4.22}$$

$M_{\cdot p}^{(P|C_k)}$ denotes the sum of transition probability of other pivotal objects reaching p in cluster C_k, which represents the importance of object p to some extent.

When considering the homogeneous relations (if existing), the probability of pivotal object p is generated by the heterogeneous and homogeneous relations, so it can be calculated as follows:

$$P(X_p^{(P)}|C_k) = Pr(X_p^{(P)}|C_k) \times M_{\cdot p}^{(P|C_k)}. \tag{4.23}$$

4.2.3.3 Posterior Probability for Objects

In order to determine the membership of objects, we need to estimate posterior probability of objects. In each subnetwork, there are two kinds of objects (i.e., pivotal and supportive objects). Because pivotal objects are the hub of subnetwork that integrate supportive objects and contain complete semantic information, we first estimate the posterior probability of pivotal objects, and then, the posterior probability of supportive objects is decided by that of pivotal objects.

Now, we consider how to estimate the posterior probability of pivotal objects $P(C_k|X^{(P)})$. According to the Bayesian rule, $P(C_k|X^{(P)}) \propto P(X^{(P)}|C_k) \times P(C_k)$. Since the cluster size $P(C_k)$ is unknown, we need to estimate an appropriate $P(C_k)$ to balance the cluster size. We use the $P(C_k)$ that maximizes the likelihood of generating pivotal objects in different clusters. The likelihood of pivotal objects is defined as:

$$logL = \sum_{p \in X^{(P)}} log[\sum_{k=1}^{K} P(X_p^{(P)}|C_k) \times P(C_k)]. \tag{4.24}$$

An EM algorithm can be utilized for the latent $P(C_k)$ by maximizing the $logL$. We can derive the Eqs. 4.25 and 4.26. Initially, we set the $P(C_k)$ with even values and then repeat the E step (i.e., Eq. 4.25) and M step (i.e., Eq. 4.26) to iteratively update the latent cluster probability until the $P(C_k)$ obtains convergence.

$$P^t(C_k|X^{(P)}) \propto P(X^{(P)}|C_k) \times P(C_k) \tag{4.25}$$

$$P^{t+1}(C_k) = \sum_{p \in X^{(P)}} P^t(C_k|X_p^{(P)}) \times \frac{1}{|X^{(P)}|} \tag{4.26}$$

Next, we estimate the posterior of supportive objects. The basic idea is that the posterior probability of supportive objects comes from its pivotal neighborhoods. We define it as follows:

$$P(C_k|X_q^{(S^J)}) = \sum_{p \in N(q)} P(C_k|X_p^{(P)}) \times \frac{1}{|N(q)|} \tag{4.27}$$

where $P(C_k|X_q^{(S^J)})$ is the probabilities of supportive object $X_q^{(S^J)}$ belonging to cluster C_k; $N(q)$ is the neighbor set of supportive object q. It means that the posterior probability of supportive object $X_q^{(S^J)}$ is the average value of its pivotal neighborhoods.

4.2.3.4   Ranking for Objects

Since the probability model obeys the PageRank principle, we can regard the conditional probability of objects as their ranks.

$$Rank(X^{(J)}) \approx P(X^{(J)}|C_k) \tag{4.28}$$

Because the conditional probability $P(X^{(J)}|C_k)$ in HeProjI is estimated by the random walk process, it may prefer to assign a higher probability to an object with a higher degree. However, in some applications, the link number-based measure is not proper. For example, advertisement webpage may have many poor value links (i.e., high degree but low rank).

If we know the additional information of objects, which can be used to measure the importance of objects, we can integrate the information into the proposed method and then get the more reasonable rank. Based on the conditional probability of objects, we propose a general ranking method for objects as follows:

$$Rank(X^{(J)}) = AI(X^{(J)}) \times P(X^{(J)}|C_k) \tag{4.29}$$

where the $AI(X^{(J)})$ is the additional importance measure (AI) of objects $X^{(J)}$. For example, in bibliographic network, the importance of a paper is decided by its citations to a large extent, and the AI can be a measure that is proportion to citations. We can also propagate the AI information to adjacent objects by transition probability matrix. It is denoted as follows:

$$Rank(X^{(I)}|C_k) = Rank(X^{(J)}|C_k) \times M^{(J,I)}. \tag{4.30}$$

4.2.4 Experiments

In this section, we evaluate the effectiveness of HeProjI and compare it with several state-of-art methods on three real datasets. In experiments, we use two real information networks: DBLP and SLAP. The schemas of these two networks are shown in Fig. 4.10c and d. In addition, we extract two different-scaled subsets of the DBLP which are called DBLP-S and DBLP-L, respectively. The DBLP-S is a small-size dataset which includes three research areas: database (DB), data mining (DM), and information retrieval (IR). While the DBLP-L is a large dataset which includes eight areas.

4.2.4.1 Clustering Effectiveness Study

In this section, we study the clustering effectiveness of HeProjI through comparing it with other well-established algorithms.

The first experiment is done on DBLP dataset, since this dataset has a relatively simple structure and is suitable for comparison with previous algorithms. The representative algorithms are included in experiments, which are summarized as follows:

- HeProjI. It is the proposed algorithm.
- HeProjI$_{\backslash s}$. It is HeProjI without considering the smooth information from general network (i.e., λ is 1 in Eq. 4.21).
- HeProjI$_{\backslash I}$. It is HeProjI without considering inheriting information from other subnetworks (i.e., Θ is 0 in Eq. 4.16).
- ComClus [27]. It is a ranking-based clustering method designed for the star-schema network with self loop.
- NetClus [22]. It is a ranking-based clustering method designed for the star-schema network without self loop.
- iTopicModel [20]. It integrates topic model and heterogeneous link information, so it can be used to do clustering in HIN.
- NetPLSA [9]. It regularizes a statistical topic model with a harmonic regularizer based on a graph structure.

The clustering quality is measured by the fraction of vertices identified correctly, FVIC [11, 15], which evaluates the average matching degree by comparing each predicting cluster with the most matching real cluster. The larger the FVIC is, the better the partition is. HeProjI, ComClus, and NetClus can be applied to DBLP dataset directly. For NetClus, we do not consider the self loop of type P, since

Table 4.7 Clustering accuracy for DBLP dataset

Accuracy		Paper (DBLP-S)	Venue (DBLP-S)	Author (DBLP-S)	Paper (DBLP-L)
HeProjI	Mean/Dev.	**0.857**/0.043	**0.823**/0.047	**0.725**/0.034	**0.603**/0.071
HeProjI$_{\backslash S}$	Mean/Dev.	0.781/0.077	0.753/0.069	0.698/0.057	0.566/0.113
HeProjI$_{\backslash I}$	Mean/Dev.	0.703/0.053	0.681/0.045	0.605/0.039	0.507/0.083
ComClus	Mean/Dev.	0.764/0.020	0.775/0.027	0.690/0.015	0.576/0.024
NetClus	Mean/Dev.	0.742/0.063	0.718/0.065	0.689/0.051	0.566/0.104
iTopicModel	Mean/Dev.	0.512/0.072	0.762/0.094	0.587/0.073	0.361/0.167
NetPLSA	Mean/Dev.	0.466/0.047	0.565/0.081	0.316/0.023	0.338/0.092

NetClus cannot solve it. Note that RankClus [21] is not included here, because it only solves the bipartite network. Moreover, for iTopicModel and NetPLSA, we make a homogeneity assumption of links so that it can be applied to this dataset. The smoothing parameter λ in HeProjI is fixed at 0.9. All learning rate Θ are fixed at 0.3. In HeProjI, the projection sequence is $P - A - C - T$. The parameters in other algorithms are set with the suggested values in their literals.

From the results shown in Table 4.7, we can observe that HeProjI achieves the best accuracy and lower standard deviation on all objects. HeProjI$_{\backslash S}$ also has good performances. However, due to omitting the smoothing operation, it has worse performances and stability when compared to HeProjI. The performances of HeProjI$_{\backslash I}$ degrade greatly, since it does not inherit clustering information from other subnetworks. In this condition, HeProjI$_{\backslash I}$ analyzes these subnetworks independently, so the inconsistency among subnetworks causes its bad performances. NetClus and ComClus both have respectable results. However, the absence of citation information among papers may lead to NetClus's worse performances when it is compared with ComClus. The iTopicModel and NetPLSA methods ignore the heterogeneity of objects and relations, so their performances are bad.

For SLAP network, contemporary methods cannot solve it directly. In order to compare with other algorithms, we convert the SLAP network into a homogeneous network through ignoring the heterogeneity of objects. As a comparison algorithm, the classical spectral clustering algorithm, NCut [19], is run on the homogeneous network. The projection sequence is $GO - G - C - T - Sub - Si$. HeProjI uses the same parameters with the above experiments, except the learning rate $\Theta[\theta_{G,GO}, \theta_{GO,G}, \theta_{G,C}, \theta_{G,T}, \theta_{C,Sub}, \theta_{C,Si}] = [0.3, 0.5, 0.7, 0.7, 0.7, 0.7]$. The results are shown in Table 4.8. It is clear that HeProjI performs much better than NCut. We know that there are distinct differences on different types of objects and relations, e.g., 70,672 links in $G - C$ relation and 2222 links in $G - GO$ relation. If we do not consider object types, as NCut does, the clusters may be seriously unbalanced, which results in the bad performances of NCut.

Table 4.8 Clustering accuracy for SLAP dataset

Accuracy	HeProjI		NCut	
	Mean	Dev.	Mean	Dev.
Gene	**0.68**	0.057	0.355	0.165
Chemical compound	**0.437**	0.031	0.307	0.091
Gene ontology	**0.557**	0.026	0.261	0.088
Tissue	**0.407**	0.066	0.293	0.09
Side effect	**0.548**	0.098	0.25	0.056
Substructure	**0.481**	0.053	0.314	0.102

4.2.4.2   Ranking Effectiveness Study

To evaluate the ranking effectiveness of HeProjI, we make a ranking accuracy comparison between HeProjI and NetClus. We utilize the venues rank recommended by Microsoft Academic Search [10] as the ground truth. In order to measure the quality of the ranking result, we employ the *Distance* criterion proposed in [12], which computes the differences between two ranking lists of the same set of objects. The criterion not only measures the number of mismatches between two lists but also gives a big penalty term to top mismatch objects in the lists. The smaller *Distance* means the better performance.

Three algorithms are tested on the DBLP dataset. In addition to NetClus, there are two versions of HeProjI (HeProjI with/without AI). The citations of paper are used as the AI measure. We extract the top 5 and 10 venues in different research areas and then calculate the *Distance* measure for them. Additionally, we also compare the accuracy of the global rank on both HeProjI and NetClus. The comparison results are shown in Fig. 4.13. We can find that two versions of HeProjI achieve better rank performances compared with NetClus in the most cases, since their *Distance* get lower values. Moreover, the HeProjI-AI performs better than HeProjI. In DBLP

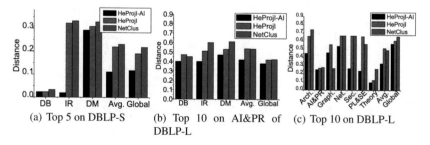

(a) Top 5 on DBLP-S   (b) Top 10 on AI&PR of   (c) Top 10 on DBLP-L
                              DBLP-L

Fig. 4.13 Ranking accuracy comparison on top venues (the smaller *Distance*, the better performance)

dataset, the citation information of papers (i.e., AI) reflects the quality of the papers to a large extent. So integrating the AI in HeProjI helps to improve the rank accuracy of papers. Moreover, the citation information can also promote the ranking accuracy of venues through the $P - V$ relation (see Eq. 4.30). So HeProjI-AI achieves the best ranking performances.

4.2.4.3 Case Study

We compare the ranking effectiveness of HeProjI and NetClus with a case study on DBLP dataset. We use the global rank to prove the ranking effectiveness of the HeProjI method. Table 4.9 shows the top 15 venues ranked by HeProjI and NetClus on DBLP-S. From these results, the ranks of venues generated by HeProjI-AI more conform to the intuition. Although it is hard to rank conferences across different areas, the order within each area is more or less established, and the HeProjI-AI confirms with that order. For example, in the DB area, it is SIGMOD, VLDB, and ICDE, while in the data mining area, it is KDD, ICDM, and PKDD. However, there are some out of order venues generated by NetClus. For example, among the database conferences, SIGMOD is ranked after VLDB and ICDE. Because NetClus cannot combine additional AI information (i.e., the citations of papers) and tends to get the rank which is proportion to its link number, it has the tendency to rank a good venue publishing a smaller number of papers with a lower rank (e.g., PODS) and a venue publishing a larger number of papers with higher rank (e.g., DEXA). Besides, for HeProjI which does not consider AI information, the rank of venues is basically proportional to their links, since the probability of objects is generated by a random walk-based method. The experiments reflect that the HeProjI method can flexibly

Table 4.9 Top 15 venues in 3 clusters on DBLP-S

Rank		1	2	3	4	5	6	7	8
HeProjI - AI	Venue	SIGMOD	VLDB	SIGIR	ICDE	KDD	PODS	WWW	CIKM
	#Papers	2428	2444	2509	2832	1531	940	1501	2204
HeProjI	Venue	ICDE	SIGIR	VLDB	SIGMOD	CIKM	DEXA	KDD	WWW
	#Papers	2832	2509	2444	2428	2204	1731	1531	1501
NetClus	Venue	VLDB	ICDE	SIGMOD	SIGIR	KDD	WWW	CIKM	ICDM
	#Papers	2444	2832	2428	2509	1531	1510	2204	1436
Rank		9	10	11	12	13	14	15	...
HeProjI-AI	Venue	ICDM	EDBT	PKDD	WSDM	PAKDD	DEXA	WebDB	
	#Papers	1436	747	680	198	1030	1731	972	...
HeProjI	Venue	ICDM	PAKDD	PODS	EDBT	PKDD	ECIR	WSDM	
	#Papers	1436	1030	1436	747	680	575	198	...
NetClus	Venue	PODS	DEXA	PAKDD	EDBT	PKDD	WSDM	ECIR	
	#Papers	940	1731	1030	747	680	198	575	...

and effectively integrate heterogeneous information and achieve more reasonable ranks. The detailed method description and validation experiments can been seen in [17].

4.3   Conclusions

Meta path is an unique characteristic of heterogeneous information network. It is an effective semantic capture tool, as well as feature extraction method. As a consequence, meta path play a critical role in data mining tasks on heterogeneous information network. In this chapter, we present two examples on ranking and clustering, respectively. Particularly, we study the ranking problem in heterogeneous information network and propose the HRank framework, which is a path-based random walk method. In addition, we study the ranking-based clustering problem in a general heterogeneous information network and proposed a novel algorithm HeProjI which projects a general HIN with arbitrary schema into a sequence of projected subnetworks and iteratively analyzes each subnetwork. Experiments not only validate their effectiveness but also illustrate the unique advantages of meta path.

Some interesting future works are worth being exploited on meta paths. On the one hand, meta path can be employed on other data mining tasks, so that we can obverse its power and potential on more applications. On the other hand, we need to design more powerful tools, beyond meta path, to capture subtle semantic.

References

1. Brin, S., Page, L.: The anatomy of a large-scale hyper textual web search engine. Comput. Netw. ISDN Syst. **30**(1–7), 1757–1771 (1998)
2. Chen, B., Ding, Y., Wild, D.: Assessing drug target association using semantic linked data. PLoS Comput. Biol. **8**(7)(e1002574), 1757–1771 (2012)
3. Grčar, M., Trdin, N., Lavrač, N.: A methodology for mining document-enriched heterogeneous information networks. Comput. J. **56**(3), 107–121 (2011)
4. Han, J.: Mining heterogeneous information networks by exploring the power of links. In: DS, pp. 13–30 (2009)
5. Jeh, G., Widom, J.: SimRank: a measure of structural-context similarity. In: KDD, pp. 538–543 (2002)
6. Ji, M., Sun, Y., Danilevsky, M., Han, J., Gao, J.: Graph regularized transductive classification on heterogeneous information networks. In: ECML/PKDD, pp. 570–586 (2010)
7. Kleinberg, J.M.: Authoritative sources in a hyperlinked environment. In: SODA, pp. 668–677 (1999)
8. Kong, X., Yu, P.S., Ding, Y., Wild, D.J.: Meta path-based collective classification in heterogeneous information networks. In: CIKM, pp. 1567–1571 (2012)
9. Mei, Q., Cai, D., Zhang, D., Zhai, C.: Topic modeling with network regularization. In: WWW, pp. 101–110 (2008)
10. Microsoft: Microsoft Academic. http://academic.research.microsoft.com
11. Newman, M.E.J., Girvan, M.: Finding and evaluating community structure in networks. Phys. Rev. E **69**(026113), 1757–1771 (2004)

12. Nie, Z., Zhang, Y., Wen, J.R., Ma, W.Y.: Object-level ranking: bringing order to web objects. In: WWW, pp. 567–574 (2005)
13. Page, L., Brin, S., Motwani, R., Winograd, T.: The pagerank citation ranking: bringing order to the web. In: Stanford InfoLab, pp. 1–14 (1998)
14. Shi, C., Kong, X., Yu, P.S., Xie, S., Wu, B.: Relevance search in heterogeneous networks. In: International Conference on Extending Database Technology, pp. 180–191 (2012)
15. Shi, C., Yan, Z., Cai, Y., Wu, B.: Multi-objective community detection in complex networks. Appl. Soft Comput. **12**(2), 850–859 (2012)
16. Shi, C., Zhou, C., Kong, X., Yu, P.S., Liu, G., Wang, B.: HeteRecom: a semantic-based recommendation system in heterogeneous networks. In: KDD, pp. 1552–1555 (2012)
17. Shi, C., Wang, R., Li, Y., Yu, P.S., Wu, B.: Ranking-based clustering on general heterogeneous information networks by network projection. In: CIKM, pp. 699–708 (2014)
18. Shi, C., Li, Y., Yu, P.S., Wu, B.: Constrained-meta-path-based ranking in heterogeneous information network. Knowl. Inf. Syst. **49**(2), 1–29 (2016)
19. Shi, J., Malik, J.: Normalized cuts and image segmentation. IEEE Trans. Pattern Anal. Mach. Intell. **22**(8), 888–905 (2000)
20. Sun, Y., Han, J., Gao, J., Yu, Y.: itopicmodel: information network-integrated topic modeling. In: ICDM, pp. 493–502 (2009)
21. Sun, Y., Han, J., Zhao, P., Yin, Z., Cheng, H., Wu, T.: RankClus: integrating clustering with ranking for heterogeneous information network analysis. In: EDBT, pp. 565–576 (2009)
22. Sun, Y., Yu, Y., Han, J.: Ranking-based clustering of heterogeneous information networks with star network schema. In: KDD, pp. 797–806 (2009)
23. Sun, Y., Norick, B., Han, J., Yan, X., Yu, P.S., Yu, X.: Integrating meta-Path selection with user-guided object clustering in heterogeneous information networks. In: KDD, pp. 1348–1356 (2012)
24. Sun, Y., Norick, B., Han, J., Yan, X., Yu, P.S., Yu, X.: Pathselclus: integrating meta-path selection with user-guided object clustering in heterogeneous information networks. ACM Trans. Knowl. Discov. Data **7**(3), 723–724 (2012)
25. Sun, Y.Z., Han, J.W., Yan, X.F., Yu, P.S., Wu, T.: PathSim: meta path-based top-K similarity search in heterogeneous information networks. In: VLDB, pp. 992–1003 (2011)
26. Tang, J., Zhang, J., Yao, L., Li, J., Zhang, L., Su, Z.: ArnetMiner: extraction and mining of academic social networks. In: KDD, pp. 990–998 (2008)
27. Wang, R., Shi, C., Yu, P.S., Wu, B.: Integrating clustering and ranking on hybrid heterogeneous information network. In: PAKDD, pp. 583–594 (2013)
28. Zhou, D., Orshanskiy, S.A., Zha, H., Giles, C.L.: Co-ranking authors and documents in a heterogeneous network. In: ICDM, pp. 739–744 (2007)

Chapter 5
Recommendation with Heterogeneous Information

Abstract Recently, heterogeneous information network (HIN) analysis has attracted a lot of attention, and many data mining tasks have been exploited on HIN. As an important data mining task, recommender system includes a lot of object types (e.g., users, movies, actors, and interest groups in movie recommendation) and the rich relations among object types, which naturally constitute an HIN. The comprehensive information integration and rich semantic information of HIN make it promising to generate better recommendation. In this chapter, we introduce three works on recommendation with HIN. One work recommends items with semantic meta paths, and the other two works extend traditional matrix factorization with rich heterogeneous information.

5.1 Recommendation Based on Semantic Path

5.1.1 Overview

In recent years, some works [5, 9, 24] have taken notice of the benefits of HIN for recommendation, where the objects and their relations in recommended system constitute a heterogeneous information network (HIN). Figure 5.1 shows such an example. The HIN not only contains different types of objects in movie recommendation (e.g., users and movies) but also illustrates all kinds of relations among objects, such as viewing information, social relations, and attribute information. Constructing heterogeneous networks for recommendation can effectively integrate all kinds of informations, which can be potentially utilized for recommendation. Moreover, the objects and relations in the networks have different semantics, which can be explored to reveal subtle relations among objects. For example, the meta path "User-Movie-User" in Fig. 5.1 means users viewing the same movies and can be used to find the similar users according to viewing records. If we recommend movies following this meta path, it will recommend the movies that are seen by users having the same viewing records with the given user. It corresponds to the collaborative filtering model in essence. Similarly, the "User-Interest Group-User" path can find the similar users with similar interests. This path corresponds to the member recommendation [25].

© Springer International Publishing AG 2017
C. Shi and P.S. Yu, *Heterogeneous Information Network Analysis
and Applications*, Data Analytics, DOI 10.1007/978-3-319-56212-4_5

Fig. 5.1 The objects and relations in movie recommended system are organized as a weighted heterogeneous information network

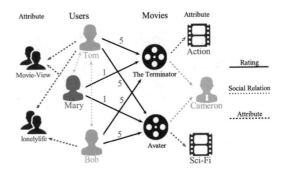

So we can directly recommend items based on the similar users generated by different meta paths connecting users. Moreover, it can realize different recommendation models through properly setting meta paths. However, this idea faces the following two challenges.

Firstly, conventional HIN and meta path cannot be directly applied to recommended system. As we know, conventional HIN and meta path do not consider the attribute values on links. However, this movie recommendation network can contain attribute values on links. Concretely, in recommended system, the users can provide a rating score to each movie viewed. The rating scores usually range from 1 to 5 as indicated on the link between user and movie in Fig. 5.1, where higher score means stronger preference. Ignoring the rating scores may result in bad similarity discovery on users. For example, according to the path "User-Movie-User," Tom has the same similarity with Mary and Bob, since they view the same movies. However, they may have totally different tastes due to different rating scores. In fact, Tom and Bob should be more similar, since they both like the same movies very much with high scores. Mary may have totally different tastes, because she does not like these movies at all. The conventional meta path does not allow links to have attribute values (e.g., rating scores in the above example) [19, 24], and hence, it cannot reveal this subtle difference. However, this difference is very important, especially in recommended system, to more accurately reveal relations of objects. So we need to extend existing HIN and meta path for considering attribute values on links. Moreover, the new similarity measures are urgently needed for development.

Secondly, it is difficult to effectively combine information from multiple meta paths for recommendation. As we have said, different types of similar users will be generated through different meta paths, and these different types of similar users will recommend different items. A weight learning method can be designed to combine these recommendations, and each path can be assigned with a learned weight preference. A good weight learning method should obtain prioritized and personalized weights. That is, the learned weights can represent the importance of paths, and each user should have personalized weights to embody his preferences on paths. The prioritized and personalized weights are very important for recommendation, since they can deeply reveal the characteristics of users. Much more than this, it makes the recommendation more explainable, since meta paths contain semantics.

For example, if a user has high-weight preference on the "User-Interest Group-User" path, we can explain that the recommendation results stem from movies viewed by users in the interest groups he joined in. Unfortunately, the personalized weights may suffer from the rating sparsity problem, especially for users with little rating information. The reasons lie in that so many parameters are needed to be learned and rating information is usually not sufficient.

In this chapter, we extend HIN and meta path for widely existing attribute values on links in information networks and, firstly, propose the weighted HIN and weighted meta path concepts to more subtly reveal object relations through distinguishing link attribute values. Instead of designing an ad hoc similarity measure for weighted meta paths, we design a novel similarity computation strategy that can make existing path-based similarity measures still usable. Furthermore, the semantic path-based personalized recommendation method SemRec is proposed to flexibly integrate heterogeneous information through setting meta paths. In SemRec, we design a novel weight regularization term to obtain personalized weight preferences on paths and alleviate the rating sparsity through employing the consistency rule of weight preferences of similar users.

5.1.2 Heterogeneous Network Framework for Recommendation

In this section, we describe notations used in this chapter and present some preliminary knowledge.

5.1.2.1 Basic Concepts

An HIN is a special type of information network with the underneath data structure as a directed graph, which contains either multiple types of objects or multiple types of links. Traditionally, HIN does not consider the attribute values on links. However, many real networks contain attribute values on links. For example, users usually rate movies with a score from 1 to 5 in movie recommended system, and the "author of" relations between authors and papers in bibliographic networks can take values (e.g., 1, 2, 3) which means the order of authors in the paper. In this chapter, we formally propose the weighted heterogeneous information network concept to handle this condition.

Definition 5.1 (*Weighted information network*) Given a schema $S = (A, R, W)$ which consists of a set of object types $A = \{A\}$, a set of relations connecting object pairs $R = \{R\}$, and a set of attribute values on relations $W = \{W\}$, a **weighted information network** is defined as a directed graph $G = (V, E, W)$ with an object type mapping function $\varphi : V \rightarrow A$, a link type mapping function $\psi : E \rightarrow R$, and an attribute value type mapping function $\theta : W \rightarrow W$. Each object $v \in V$ belongs to

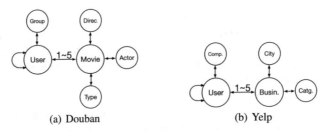

Fig. 5.2 Network schema of weighted heterogeneous information networks constituted by two datasets

one particular object type $\varphi(v) \in A$, each link $e \in E$ belongs to a particular relation $\psi(e) \in R$, and each attribute value $w \in W$ belongs to a particular attribute value type $\theta(w) \in W$. When the types of objects $|A| = 1$ and the types of relations $|R| = 1$, it is a **homogeneous information network**. When the types of objects $|A| > 1$ (or the types of relations $|R| > 1$) and the types of attribute values $|W| = 0$, the network is called **unweighted heterogeneous information network**. When the types of objects $|A| > 1$ (or the types of relations $|R| > 1$) and the types of attribute values $|W| > 0$, the network is called **weighted heterogeneous information network** (WHIN).

Conventional HIN is an unweighted HIN, where there are no attribute values on relations or we do not consider them. For a WHIN, there are attribute values on some relation types, and these attribute values may be discrete or continuous values.

Example 5.1 A movie recommended system can be organized as a weighted heterogeneous information network, whose network schema is shown in Fig. 5.2a. The network contains objects from six types of entities (e.g., users, movies, groups, actors) and relations between them. Links between objects represent different relations. For example, links exist between users and users denoting the friendship relations, between users and movies denoting rating and rated relations. In addition, the network also contains one type of attribute value on the rating relation between users and movies, which take values from 1 to 5.

Two objects in an HIN can be connected via different paths, and these paths have different meanings. As an example shown in Fig. 5.2a, users can be connected via "User-User" (UU) path, "User-Group-User" (UGU) path, "User-Movie-User" (UMU), and so on. These paths are called meta paths that are the combination of a sequence of relations between object types. Although meta path is widely used to reveal semantics among objects [20], it fails to distinguish the attribute values between two objects in WHIN. For example, if ignoring the different rating scores of users on items in above movie recommendation, we may obtain incorrect results. Consider a scenario that we use the UMU path to find the similar users of Tom according to their viewing records in Fig. 5.1. We can infer that Tom is very similar to Mary and Bob, since they have the same viewing records. However, it is obvious that Tom and Mary have totally different tastes. So the UMU path cannot subtly reveal the different ratings of users on the same movies. In order to effectively exploit semantics in WHIN, we extend the conventional meta path to consider attribute values on relations. Without loss of generality, we assume the attribute values on relations in WHIN are discrete. For continuous attribute values on relations, we can convert the continuous attribute values into discrete ones.

Definition 5.2 (*Extended meta path on WHIN*) Extended meta path is a meta path based on a certain attribute value constraint on relations, which is denoted as $A_1 \xrightarrow{\delta_1(R_1)} A_2 \xrightarrow{\delta_2(R_2)} \xrightarrow{\delta_l(R_l)} A_{l+1}|C$ (also denoted as $A_1(\delta_1(R_1))A_2(\delta_2(R_2)) \cdots \cdots (\delta_l(R_l))A_{l+1}|C$). If the relation R has attribute values on links, the attribute value function $\delta(R)$ is a set of values from the attribute value range of relation R, else $\delta(R)$ is an empty set. $A_i \xrightarrow{\delta_i(R_i)} A_{i+1}$ represents the relation R_i between A_i and A_{i+1} based on the attribute values $\delta_i(R_i)$. The constraint C on attribute value functions is a set of correlation constraints among attribute value functions. If all attribute value functions in a meta path are empty set (the corresponding constraint C is also an empty set), the path is called an **unweighted meta path**, else the path is called a **weighted meta path**.

Note that the conventional meta path is an unweighted meta path that can be considered as the special case of a weighted meta path.

Example 5.2 Taking Fig. 5.2a as an example, the rating relation between users U and movies M can take scores from 1 to 5. The weighted meta path $U \xrightarrow{1} M$ (i.e., $U(1)M$) means movies rated by users with score 1, which implies that users dislike the movies. The weighted meta path $U \xrightarrow{1,2} M \xrightarrow{1,2} U$ (i.e., $U(1, 2)M(1, 2)U$) means users disliking the same movies as the target user, while the unweighted meta path UMU can only reflect that users have the same viewing records. Furthermore, we can flexibly set the correlation constraints of attribute value functions on different relations in weighted meta paths. For example, the path $U(i)M(j)U|i = j$ means users having exactly the same ratings on some movies as the target user. Under this path, we can easily find that, in Fig. 5.1, Tom is very similar to Bob, while they are totally dissimilar to Mary.

5.1.2.2 Recommendation on Heterogeneous Networks

For a target user, recommended systems usually recommend items according to his similar users. In HIN, there are a number of meta paths connecting users, such as "User-User" and "User-Movie-User". Based on these paths, users have different similarities. Here, we define the path-based similarity as follows.

Definition 5.3 (*Path-based similarity*) In HIN, the path-based similarity of two objects is the similarity evaluation based on the given meta path connecting these two objects.

Table 5.1 The meanings and corresponding recommendation models of meta paths

No.	Meta path	Semantic meaning	Recommendation model
1	UU	Friends of the target user	Social recommendation
2	UGU	Users in the same group of the target user	Member recommendation
3	UMU	Users who view the same movies with the target user	Collaborative recommendation
4	UMTMU	Users who view the movies having the same types with that of the target user	Content recommendation

After obtaining the path-based similarity of users, we can recommend items according to the similar users of the target user. More importantly, the meta paths connecting users have different semantics, which can represent different recommendation models. As an example shown in Fig. 5.2a, "User-User" (UU) means friends of the target user. If we recommend movies according to the similarity of users generated by that path, it will recommend the movies viewed by friends of the target user. Indeed, it is the social recommendation. Another example is that "User-Movie-User" (UMU) means users who view the same movies with the target user. Following that path, it will recommend the movies viewed by users having the similar viewing records with the target user. It is collaborative recommendation in essential. Table 5.1 shows the other representative paths and the corresponding recommendation models. Based on the HIN framework, we can flexibly represent different recommendation models through properly setting meta paths.

5.1.2.3   Similarity Measure Based on Weighted Meta Path

Similarity measure on meta paths have been well studied, and several path-based similarity measures have been proposed on HIN, such as PathSim [19], PCRW [6], and HeteSim [16]. However, these similarity measures cannot be directly applied to weighted meta path, because they do not consider the attribute value constraint on relations. As we know, the essential of the path-based similarity measure is to evaluate the proportion of the number of paths connecting two objects on all possible paths along the meta path [19], so the paths along a weighted meta path must satisfy the attribute value constraint. Moreover, the attribute value on relations may be a variable, even correlated. Taking the $U(i)M(j)U|i = j$ path as an example, the attribute values i and j are variables from 1 to 5, and they satisfy constraint $i = j$. For this kind of paths, existing path-based similarity measures cannot handle it.

In order to address the variable, even correlated, attribute value constraints in a weighted meta path, we extend the meta path concept and propose a general strategy to make existing path-based similarity measure still usable, instead of proposing an ad hoc similarity measure. Specifically, we can decompose the weighted meta path into a group of atomic meta paths with fixed attribute value constraint. For an atomic meta path, the existing path-based similarity measures can be used directly.

Definition 5.4 (*Atomic meta path*) If all attribute value functions $\delta(R)$ in a weighted meta path take a specific value, the path is called an **atomic meta path**. A weighted meta path is **a group of atomic meta paths** which contain all atomic meta paths that satisfy the constraint C.

Example 5.3 Taking Fig. 5.2a as an example, $U(1)M(1)U$ and $U(1)M(2)U$ both are atomic meta paths. The weighted meta path $U(i)M(j)U|i = j$ is a group of five atomic meta paths (e.g., $U(1)M(1)U$ and $U(2)M(2)U$).

Since a weighted meta path is a group of corresponding atomic meta paths, the similarity measure based on a weighted meta path can be considered as the sum of the similarity measure based on the corresponding atomic meta paths. So the similarity measure based on a weighted meta path can be evaluated based on the following two steps: (1) Evaluate the similarity based on each atomic meta path with existing path-based measures; (2) sum up the similarities on all atomic meta paths in the weighted meta path. Note that the similarity measure needs to consider the effect of the normalized term existing in some path-based similarity measures, such as PathSim [19] and HeteSim [16]. Taking PathSim as an example, we illustrate its calculation process along conventional and weighted meta path in Fig. 5.3, where the rating matrix between 3 users and 2 movies is from Fig. 5.1. We know that PathSim counts the number of path instances connecting two objects along conventional meta path with a normalized term (shown in the upper half of Fig. 5.3), and thus, it regards that the users all are the same. As shown in the lower half of Fig. 5.3, PathSim along weighted meta path firstly counts the number of path instances along each atomic meta path and then sums up the number of path instances along all atomic meta paths before normalization. And thus, it can more accurately discover that only u_1 and u_3 are similar, since they have the same tastes in movies.

5.1.3 The SemRec Solution

In this section, we proposed a **Sem**antic path-based personalized **Rec**ommendation method (**SemRec**) to predict the scores of items. Specifically, SemRec first evaluates the similarity of users based on weighted or unweighted meta paths and then infers the predicted scores on items according to the rating scores of similar users. Under different meta paths, the users can obtain different recommendation results. How to effectively combine these recommendations generated by different meta paths is challenging. We need to put different preferences on the various meta paths. This

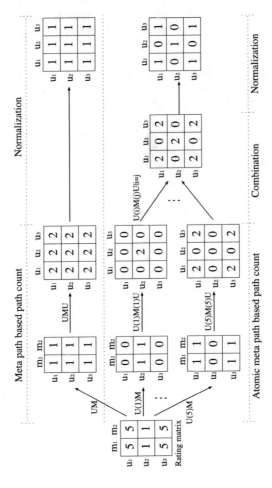

Fig. 5.3 PathSim similarity measure based on conventional and weighted meta path

results in assigning preference weight to each meta path. We abbreviate the preference weight as weight when the context is clear without confusion with the link weight in the weighted meta path. There are two aspects of difficulties in learning the weights. (1) Prioritized weights: That is, the weights learned should embody the importance of paths and reflect users' preferences. However, the similarity evaluations based on different paths have significant bias, which makes path preference hard to reflect the path importances. For example, the similarity evaluations may all be high based on a path with dense relations, while the similarity evaluations may all be low based on another path with sparse relations. So the similarity evaluations based on different paths cannot reflect the similarity of two objects. SemRec designs a normalized rating intensity operation to eliminate the similarity bias, which makes the weight better reflect path importances. (2) Personalized weights: That is, it is better to learn weight preferences for each user. However, personalized weight learning may suffer from the rating sparsity problem, since many users have little rating informations. In order to alleviate the rating sparsity problem for personalized weight learning, we propose *the consistency rule of weight preferences of similar users*. That is, we assume that two similar users have consistent weight preferences on meta paths. While it is reasonable, it is seldom used before. Two users are similar based on a path, which implies the path has similar impacts on these two users. That is to say, these users have the consistent preferences on the path. Following this principle, we design a novel weight regularization term, which effectively alleviates rating sparsity in personalized weight learning.

In the following sections, we firstly design the basic recommendation method based on a single path. And then, we propose three levels of personalized recommendation methods based on multiple paths: unified weights for all users, personalized weights for each user, and personalized weights with weight regularization.

5.1.3.1 Recommendation with Single Path

Based on the path-based similarity of users, we can find the similar users of a target user under a given path, and then, the rating score of the target user on an item can be inferred according to the rating scores of his similar users on the item. Assume that the range of rating scores are from 1 to N (e.g., 5); P is a set of unweighted or weighted meta paths; $R \in \mathbf{R}^{|U| \times |I|}$ is the rating matrix, where $R_{u,i}$ denotes the rating score of user u on item i; and $S \in \mathbf{R}^{|U| \times |U|}$ is the path-based similarity matrix of users, where $S_{u,v}^{(l)}$ is the similarity of users u and v under path P_l. Here, we define the *rating intensity* $Q \in \mathbf{R}^{|U| \times |I| \times N}$, where $Q_{u,i,r}^{(l)}$ represents the intensity of user u rating item i with score r given path P_l. $Q_{u,i,r}^{(l)}$ is determined by two aspects: the number of similar users rating the item i with score r and the similarity of users. So we calculate $Q_{u,i,r}^{(l)}$ as the sum of similarity of users rating i with r.

$$Q_{u,i,r}^{(l)} = \sum_v S_{u,v}^{(l)} \times E_{v,i,r}$$

$$E_{v,i,r} = \begin{cases} 1 & R_{v,i} = r \\ 0 & \text{others} \end{cases} \tag{5.1}$$

where $E_{v,i,r}$ indicates whether user v rates item i with score r.

Under a meta path P_l, the rating of a user u on an item i ranges from 1 to N with different rating intensities $Q_{u,i,r}^{(l)}$. So the *predicted rating score*, denoted as $\hat{R}_{u,i}^{(l)}$, of user u on item i under the path P_l can be the average of rating scores weighted by corresponding normalized intensity.

$$\hat{R}_{u,i}^{(l)} = \sum_{r=1}^N r \times \frac{Q_{u,i,r}^{(l)}}{\sum_{k=1}^N Q_{u,i,k}^{(l)}} \tag{5.2}$$

and $\hat{R}^{(l)} \in \mathbf{R}^{|U| \times |I|}$ means the predicted rating matrix under path P_l.

According to Eq. 5.2, we can predict the rating score of a user on an item under a given path and then recommend the item with the high score for a target user. Moreover, Eq. 5.2 has an additional advantage that it eliminates the similarity bias existing in different meta paths. As we know, the similarity of users under different meta paths has different scales, which makes similarity evaluation and rating intensity incomparable among different paths. The normalized rating intensity in Eq. 5.2 is able to eliminate those scale differences.

5.1.3.2 Recommendation with Multiple Paths

Under different meta paths, there are different predicted rating scores. In order to calculate the compositive score, we propose three different weight learning methods corresponding to different levels of personalized weights of users.

Unified weight learning for all users For all users, we assign each meta path with a unified weight, which means the user preference on the path. This weight vector is denoted as $w \in \mathbf{R}^{1 \times |P|}$, and $w^{(l)}$ means the weight on path P_l. The final predicted rating score under all meta paths, denoted as $\hat{R}_{u,i}$, can be the weighted sum of predicted rating score under each meta path.

$$\hat{R}_{u,i} = \sum_{l=1}^{|P|} w^{(l)} \times \hat{R}_{u,i}^{(l)} \tag{5.3}$$

Hopefully, the predicted rating matrix $\hat{R} \in \mathbf{R}^{|U| \times |I|}$ should be as close as to the real rating matrix R. So a direct optimization objective can be defined as the square error between the real scores and the predicted scores.

$$\min_{w} L_1(w) = \tfrac{1}{2}||Y \odot (R - \sum_{l=1}^{|P|} w^{(l)} \hat{R}^{(l)})||_2^2 + \tfrac{\lambda_0}{2}||w||_2^2 \tag{5.4}$$
$$s.t. \qquad w \geq 0$$

where the notation $\odot$ is the Hadamard product (also know as the entrywise product) between matrices, and $|| \cdot ||_p$ is the matrix L^p-norm. Y is an indicator matrix with $Y_{u,i} = 1$ if user u rated item i, and otherwise, $Y_{u,i} = 0$.

Personalized weight learning for individual user The above optimization objective has a basic assumption: All users have the same path preferences. However, in many real applications, each user has his personal interest preferences. Unified weights cannot provide personalized recommendations for users. To realize personalized recommendation, each user is assigned with weight vector on meta paths. The weight matrix is denoted as $W \in \mathbf{R}^{|U| \times |P|}$, in which each entry, denoted as $W_u^{(l)}$, means the preference weight of user u on path P_l. The column vector $W^{(l)} \in \mathbf{R}^{|U| \times 1}$ means the weight vector of all users on path P_l. So the predicted rating $\hat{R}_{u,i}$ of user u rating item i under all paths is as follows:

$$\hat{R}_{u,i} = \sum_{l=1}^{|P|} W_u^{(l)} \times \hat{R}_{u,i}^{(l)} \tag{5.5}$$

Similarly, we can define the optimization objective as follows:

$$\min_{W} L_2(W) = \tfrac{1}{2}||Y \odot (R - \sum_{l=1}^{|P|} diag(W^{(l)})\hat{R}^{(l)})||_2^2 + \tfrac{\lambda_0}{2}||W||_2^2 \tag{5.6}$$
$$s.t. \qquad W \geq 0$$

where $diag(W^{(l)})$ means the diagonal matrix transformed from a vector $W^{(l)}$.

Personalized weight learning with weight regularization Although Eq. 5.6 consider user's personalized weights, it may be hard to effectively learn weights for those users that have little rating information. There are $|U| \times |P|$ weight parameters to learn, while the training samples are usually much smaller than $|U| \times |I|$. The training samples are usually not sufficient for the weight learning, specially for those cold-start users and items. According to the consistency rule of weight preferences of similar users mentioned above, the path weights of a user should be consistent to that of his similar users. For users with little rating information, their path weights can be learnt from the weights of their similar users, since the similarity information of users are more available through meta paths. So we design a weight regularization term as follows, which compels the weights of a user consistent to the average of weights of his similar users.

$$\sum_{u=1}^{|U|} \sum_{l=1}^{|P|} (W_u^{(l)} - \sum_{v=1}^{|U|} \bar{S}_{u,v}^{(l)} W_v^{(l)})^2 \tag{5.7}$$

where $\bar{S}_{u,v}^{(l)} = \frac{S_{u,v}^{(l)}}{\sum_v S_{u,v}^{(l)}}$ is the normalized user similarity based on path P_l. For convenience, the weight regularization term can be written as the following matrix format:

$$\sum_{l=1}^{|P|} ||W^{(l)} - \bar{S}^{(l)} W^{(l)}||_2^2 \tag{5.8}$$

And thus, the optimization objective is defined as follows:

$$\begin{aligned}
\min_W \mathrm{L}_3(W) = \ & \tfrac{1}{2}||Y \odot (R - \sum_{l=1}^{|P|} diag(W^{(l)}) \hat{R}^{(l)})||_2^2 \\
+ \ & \tfrac{\lambda_1}{2} \sum_{l=1}^{|P|} ||W^{(l)} - \bar{S}^{(l)} W^{(l)}||_2^2 + \tfrac{\lambda_0}{2}||W||_2^2
\end{aligned} \tag{5.9}$$
$$s.t. \qquad W \geq 0$$

The above optimization objective is a nonnegative quadratic programming problem, a simple special case of nonnegative matrix factorization. Projected gradient method for nonnegative bound-constrained optimization [7] can be applied to solve this problem. The gradient of Eq. 5.9 with respect to $W_u^{(l)}$ can be calculated as follows:

$$\begin{aligned}
\frac{\partial \mathrm{L}_3(W)}{\partial W_u^{(l)}} = \ & -(Y_u \odot (R_u - \sum_{l=1}^{|P|} W_u^{(l)} \hat{R}_u^{(l)}))^T \hat{R}_u^{(l)} + \lambda_0 W_u^{(l)} \\
+ \ & \lambda_1 (W_u^{(l)} - \bar{S}_u^{(l)} W^{(l)}) - \lambda_1 \bar{S}_u^{(l)T}(W^{(l)} - \bar{S}^{(l)} W^{(l)})
\end{aligned} \tag{5.10}$$

$W_u^{(l)}$ can be updated as follows:

$$W_u^{(l)} = max(0, W_u^{(l)} - \alpha \frac{\partial \mathrm{L}_3(W)}{\partial W_u^{(l)}}) \tag{5.11}$$

where α is the step size and can be set according to [7]. Algorithm 1 shows the framework of this version of SemRec.

5.1.4   Experiments

In this section, extensive experiments on two real datasets illustrate the traits of SemRec. We first validate the effectiveness of SemRec, especially for cold-start problem. Then, we thoroughly explore the meanings of weights learned and validate the benefits of the proposed weighted meta path.

Algorithm 1 Framework of SemRec

Require:
   G: weighted heterogeneous information network
   P: meta paths connecting users
   λ_0 and λ_1: controlling parameter
   α: step size for updating parameters
   ε: convergence tolerance
Ensure:
   W: the weight matrix of all users on all paths.
1: **for** $P_l \in P$ **do**
2:    Evaluate user similarity $S^{(l)}$
3:    Calculate rating intensity $Q^{(l)}$ with Eq. 5.1
4:    Calculate predicted rating score $\hat{R}^{(l)}$ with Eq. 5.2
5: **end for**
6: Initialize $W > 0$
7: **repeat**
8:    $W_{old} := W$
9:    Calculate $\frac{\partial L_3(W)}{\partial W}$ with Eq. 5.10
10:    $W := max(0, W - \alpha \frac{\partial L_3(W)}{\partial W})$
11: **until** $|W - W_{old}| < \varepsilon$

5.1.4.1 Experiment Settings

In order to get more comprehensive heterogeneous information, we crawled a new dataset from Douban,[1] a well-known social media network in China. The dataset includes 13,367 users and 12,677 movies with 1,068,278 movie ratings ranging from 1 to 5. The dataset includes the social relation among users and the attribute information of users and movies. Another dataset is the Yelp challenge dataset.[2] This dataset contains user ratings on local business and attribute information of users and businesses. The dataset includes 16,239 users and 14,284 local businesses with 198,397 ratings from 1 to 5. The detailed description of these two datasets can be seen in Table 5.2, and their network schemas are shown in Fig. 5.2. We can find that these two datasets have different properties. The Douban dataset has dense rating relations but sparse social relations, while the Yelp dataset has sparse rating relations but dense social relations.

We use two widely used metrics, Root Mean Square Error (RMSE) and Mean Absolute Error (MAE), to measure the rating prediction quantity.

$$RMSE = \sqrt{\frac{\sum_{(u,i)\in R_{test}} (R_{u,i} - \hat{R}_{u,i})^2}{|R_{test}|}} \tag{5.12}$$

$$MAE = \frac{\sum_{(u,i)\in R_{test}} |R_{u,i} - \hat{R}_{u,i}|}{|R_{test}|} \tag{5.13}$$

---

[1] http://movie.douban.com/.

[2] http://www.yelp.com/dataset_challenge/.

Table 5.2 Statistics of Douban and Yelp datasets

Dataset	Relations (A-B)	Number of A	Number of B	Number (A-B)	Ave. degrees of A/B
Douban	User–Movie	13367	12677	1068278	79.9/84.3
	User–User	2440	2294	4085	1.7/1.8
	User–Group	13337	2753	570047	42.7/207.1
	Movie–Director	10179	2449	11276	1.1/4.6
	Movie–Actor	11718	6311	33587	2.9/5.3
	Movie–Type	12676	38	27668	2.2/728.1
Yelp	User–Business	16239	14284	198397	12.2/13.9
	User–User	10580	10580	158590	15.0/15.0
	User–Compliment	14411	11	76875	5.3/6988.6
	Business–City	14267	47	14267	1.0/303.6
	Business–Category	14180	511	40009	2.8/78.3

where $R_{u,i}$ denotes the real rating user u gave to item i and $\hat{R}_{u,i}$ denotes the predicted rating. R_{test} denotes whole test set. A smaller MAE or RMSE means a better performance.

In order to show the effectiveness of the proposed SemRec, we compare four variations of SemRec with the state of the arts. Besides the personalized weight learning method with weight regularization (called SemRec$_{Reg}$), we include three special cases of SemRec: single path-based method (called SemRec$_{Sgl}$), unified weight learning method for all users (called SemRec$_{All}$), and personalized weight learning method for individual user (called SemRec$_{Ind}$). As the baselines, four representative rating predication methods are illustrated as follows. Note that the top k recommendation methods [5, 24] are not included here, since they solve different problems.

- **PMF** [14]: It is the basic matrix factorization method using only user–item matrix for recommendations.
- **SMF** [13]: It adds the social regularization term into PMF, which aims at getting the users' latent factor closer to their friends' latent factors.
- **CMF** [8]: A collective matrix factorization method, which factorizes all relations in HIN and shares the latent factor of same object types in different relations.
- **HeteMF** [22]: A matrix factorization method with entity similarity regularization, which also utilizes the relations in HIN.

We employ 5 meaningful meta paths whose lengths are not longer than 4 for both datasets, since the longer meta paths are not meaningful and they fail to produce good similarity measures [19]. Table 5.3 shows those paths which include the weighted and unweighted meta paths. For SemRec, we use PathSim [19] as the similarity measure

Table 5.3 Meta paths used in experiments

Douban	Yelp
UGU	UU
$U(i)M(j)U \mid i = j$	UCoU
$U(i)MDM(j)U \mid i = j$	$U(i)B(j)U \mid i = j$
$U(i)MAM(j)U \mid i = j$	$U(i)BCaB(j)U \mid i = j$
$U(i)MTM(j)U \mid i = j$	$U(i)BCiB(j)U \mid i = j$

to calculate the similarity between users. The parameter λ_0 in SemRec is 0.01, and λ_1 is 10^3 for the best performance. The parameters in other methods are set with the best performances on these datasets.

5.1.4.2 Effectiveness Experiments

For Douban dataset, we use different training data settings (20%, 40%, 60%, 80%) to show the comparison results in different data sparseness. Training data 20%, for example, means that 20% of the ratings from user–item rating matrix is randomly selected as the training data to predict the remaining 80%. From Table 5.2, we can find that the Douban dataset has dense rating relations, while Yelp has very sparse rating relations. So we utilize more training data (60%, 70%, 80%, 90%) on Yelp. The random selection was repeated 10 times independently, and the average results are reported in Table 5.4. Note that SemRec$_{Sgl}$ reports the best performances on these five paths.

From the results, we can observe that all versions of SemRec outperform other approaches in most conditions. Particularly, SemRec$_{Reg}$ always achieves the best performances on all conditions. For example, on 20% training set of Douban, SemRec$_{Reg}$ outperforms PMF up to 19.55% on RSME and 15.89% on MAE. As compared to PMF, CMF improves the recommendation performances through integrating heterogeneous information with matrix factorization. However, its performances are much worse than the proposed SemRec on all conditions, especially on less training set. As the most similar method to SemRec, HeteMF also has good performances, while its performances are still worse than the proposed SemRec$_{Reg}$. These all imply that the proposed SemRec has better mechanism to integrate heterogeneous information.

In addition, different versions of SemRec have different performances. Generally, SemRec with multiple paths (e.g., SemRec$_{All}$ and SemRec$_{Reg}$) have better performances than SemRec with single path (i.e., SemRec$_{Sgl}$) except SemRec$_{Ind}$, which indicates that the weight learning of SemRec can effectively integrate the similarity information generated by different paths. Because of rating sparsity, SemRec$_{Ind}$ has worse performances than SemRec$_{All}$ on most conditions. In addition, the better performances of SemRec$_{Rec}$ over SemRec$_{Ind}$ confirm the benefit of the weight regularization term. In all, SemRec$_{Reg}$ always achieves best performances in all conditions.

Table 5.4 Effectiveness experimental results (Res. and Imp. are the abbreviations of result and improvement. The improvement is based on PMF)

Dataset	Method	Criteria	20%		40%		60%		80%		Running Time(s)
			Res.	Imp.	Res.	Imp.	Res.	Imp.	Res.	Imp.	
Douban	PMF	RMSE	0.9750		0.8455		0.7975		0.7673		260.25
		MAE	0.7198		0.6319		0.6010		0.5812		
	SMF	RMSE	0.9743	0.07%	0.8449	0.07%	0.7967	0.10%	0.7674	−0.01%	266.78
		MAE	0.7192	0.08%	0.6313	0.09%	0.6002	0.13%	0.5815	−0.05%	
	CMF	RMSE	0.9285	4.77%	0.8273	2.15%	0.8042	−0.84%	0.7741	−0.89%	509.31
		MAE	0.6971	3.15%	0.6263	0.89%	0.6090	−1.33%	0.5900	−1.51%	
	HeteMF	RMSE	0.8513	12.69%	0.7796	7.79%	0.7601	4.69%	0.7550	1.60%	736.85
		MAE	0.6342	11.89%	0.5927	6.20%	0.5800	3.49%	0.5758	0.93%	
	SemRec$_{Sgl}$	RMSE	0.8434	13.50%	0.8138	3.75%	0.7937	0.48%	0.7846	−2.25%	0
		MAE	0.6506	9.61%	0.6351	−0.51%	0.6172	−2.70%	0.6142	−5.68%	
	SemRec$_{All}$	RMSE	0.8125	16.67%	0.7814	7.58%	0.7709	3.34%	0.7656	0.22%	1.44
		MAE	0.6309	12.35%	0.6149	2.69%	0.6098	−1.46%	0.6072	−4.47%	
	SemRec$_{Ind}$	RMSE	0.8753	10.23%	0.8083	4.40%	0.7729	3.08%	0.7540	1.73%	155.98
		MAE	0.6412	10.92%	0.6032	4.54%	0.5840	2.83%	0.5739	1.26%	
	SemRec$_{Reg}$	RMSE	0.7844	19.55%	0.7452	11.86%	0.7296	8.51%	0.7216	5.96%	293.14
		MAE	0.6054	15.89%	0.5808	8.09%	0.5698	5.19%	0.5639	2.98%	

(continued)

Table 5.4 (continued)

Dataset	Method	Criteria	20%		40%		60%		80%		Running Time(s)
			Res.	Imp.	Res.	Imp.	Res.	Imp.	Res.	Imp.	
Yelp	PMF	RMSE	1.6779		1.5931		1.5323		1.4833		31.8
		MAE	1.2997		1.2262		1.1740		1.1324		
	SMF	RMSE	1.4843	11.54%	1.4017	12.01%	1.3678	10.74%	1.3377	9.82%	51.19
		MAE	1.0830	16.67%	1.0547	13.99%	1.0282	12.42%	1.0085	10.94%	
	CMF	RMSE	1.6161	3.68%	1.5731	1.26%	1.5194	0.84%	1.4793	0.27%	375.38
		MAE	1.2628	2.84%	1.2224	0.31%	1.1740	0.00%	1.1405	−0.72%	
	HeteMF	RMSE	1.2333	26.50%	1.2090	24.11%	1.1895	22.37%	1.1755	20.75%	619.25
		MAE	0.9268	28.69%	0.9107	25.73%	0.8969	23.60%	0.8878	21.60%	
	SemRec$_{Sgl}$	RMSE	1.3252	21.02%	1.2889	19.09%	1.2576	17.93%	1.2331	16.87%	0
		MAE	0.9657	25.70%	0.9420	23.18%	0.9224	21.43%	0.9067	19.93%	
	SemRec$_{All}$	RMSE	1.2166	27.49%	1.1906	25.27%	1.1665	23.87%	1.1496	22.50%	0.25
		MAE	0.9040	30.45%	0.8873	27.64%	0.8723	25.70%	0.8616	23.91%	
	SemRec$_{Ind}$	RMSE	1.3654	18.62%	1.3229	16.96%	1.2922	15.67%	1.2658	14.66%	57.22
		MAE	1.0029	22.84%	0.9728	20.67%	0.9517	18.94%	0.9322	17.68%	
	SemRec$_{Reg}$	RMSE	1.2025	28.33%	1.1760	26.18%	1.1559	24.56%	1.1423	22.99%	374.57
		MAE	0.8901	31.51%	0.8696	29.08%	0.8548	27.19%	0.8442	25.45%	

The reason lies in that SemRec$_{Reg}$ not only realizes personalized weight learning for all users but also avoids the rating sparsity through the weight regularization in it.

Furthermore, we record the average running time of these methods on the learning process. For two similarity based methods (e.g., SemRec and HeteMF), we do not consider the running time on similarity evaluation, since it can be done off-line beforehand. For the four versions of SemRec, their running times increase when the weight learning tasks become more complex. Both SemRec$_{Sgl}$ and SemRec$_{All}$ are very fast, which can be applied for online learning. The running times of SemRec$_{Ind}$ and SemRec$_{Reg}$ are still acceptable when compared to CMF and HeteMF. We can select a proper model through balancing the efficiency and effectiveness of SemRec in real applications.

5.1.4.3 Study on Cold-Start Problem

The above results also show that SemRec has more obvious superiority with less training set, which implies that SemRec has the potential to alleviate the cold-start problem. In this section, we will exploit the ability of SemRec on alleviating the cold-start problem through observing its performances on different levels of cold-start users and items. We run PMF, CMF, HeteMF, SemRec$_{Ind}$, and SemRec$_{Reg}$ on Douban dataset with users having the different numbers of rated movies. We select four types of users: three types of cold-start users with different numbers of rated movies (e.g., users with the number of rated movies no more than 5, denoted as ≤ 5 in Fig. 5.4) and all users (called ALL in Fig. 5.4). In addition, we also do the similar experiments on cold-start items and users&items (contain both cold-start users and items). We record the RMSE performance improvement of other four algorithms against PMF in Fig. 5.4.

It is clear that SemRec$_{Reg}$ always achieves the best performance improvements on almost all conditions, and its superiority is more significant for less rating information. On the contrary, CMF only achieves improvements on cold-start users and HeteMF's improvements are only on items. We think the reason lies in that the collective matrix factorization of all relations in CMF may introduce much noises, especially for items. HeteMF only utilizes the similarity information of items, ignoring

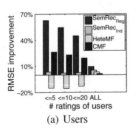 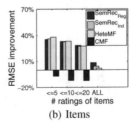 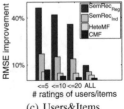

(a) Users   (b) Items   (c) Users&Items

Fig. 5.4 Performance improvements of three HIN methods against PMF on different levels and types of cold-start problems

that of users. Generally, integrating heterogeneous information is helpful in alleviating cold-start problem (see Fig. 5.4c), while the integrating mechanisms may have different impacts on cold-start items and users. The overall performance improvements of SemRec$_{Reg}$ are attributed to multiple meta paths that not only contain rich attribute information but also provide comprehensive and complementary similarity evaluation of users and items. In addition, the better performances of SemRec$_{Reg}$ over SemRec$_{Ind}$ further validate that the weight regularization term employed in SemRec$_{Reg}$ is really helpful for the weight learning of cold-start users from similar users.

5.1.4.4 Study of Weight Preferences

In this section, we illustrate the meanings of weights learned by SemRec through a case study. Based on the results of SemRec$_{Reg}$ on Douban dataset with 60% training data in the above experiments, we cluster users' weight vectors into 5 groups using K-means and then show the statistics information of users in five clusters in Fig. 5.5a. Moreover, the weight preferences of the five cluster centers on 5 meta paths are also shown in Fig. 5.5b.

Let us observe the relationship of the statistics information of users in different clusters and their weight preferences on paths from Fig. 5.5a, b. As we know, Douban is a unique social media platform in China, in which the major active users are young people who love culture and arts. As the typical and major users in Douban, the users in C3 view a good number of movies, give relatively good rating scores, and have a moderate number of friends. So they also have close weight preferences on all paths. As the top movie fans, the users in C4 view a great many movies, tend to give lower rating scores due to critical attitude, and have many friends. And they obviously like to get recommendation from viewing records of other users (i.e., UMU) and interest group (i.e., UGU), but less paying attentions to movies' content (e.g., UMTMU and

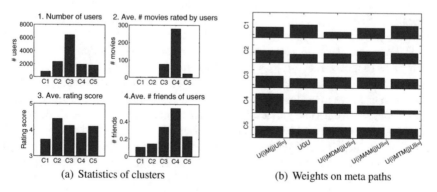

(a) Statistics of clusters   (b) Weights on meta paths

Fig. 5.5 Analysis of clusters' characteristics and path preferences of results returned by SemRec$_{Reg}$ on Douban dataset. C1–C5 represents the index of five clusters

UMAMU). In addition, the users in C1 and C2 are two types of inactive users, and they view few movies and have few friends. Because of not being fond of movies, these users tend to give much high or low rating scores. These users comparatively prefer to follow movie content (e.g., UMTMU and UMAMU). The picky users in C1 is more likely to get recommendation from interest group (i.e., UGU), while the idealess users in C2 give more preferences to viewing records of other users (i.e., UMU).

In all, the weights of paths learned by SemRec can reflect the users' path preferences, and these path preferences are able to reveal the users' characteristics to a large extent. More importantly, the meaningful weight preferences are very useful for recommendation explanation. We know that the meta path has semantics, so we can tell users the recommendation reason according to the path semantics of the high-weight path. Although some weight learning methods on paths have been proposed [9, 24], their weights fail to reflect users' preferences on paths. We think two strategies adopted in RecSem contribute to its good properties. (1) We design the predicted rating score in Eq. 5.2, which can eliminate the similarity bias on different meta paths by the adoption of normalized rating intensity. (2) We employ the weight regularization term in Eq. 5.9 according to the consistency rule of weight preferences of similar users. The consistency rule makes similar users have similar weight preferences. In other words, weights also reveal users' similarity and preferences.

5.1.4.5 Study on Weighted Meta Path

In this section, we study the effectiveness of weighted meta path on improving the performances of SemRec through more accurately revealing relations among objects. For the meta path UMU, we design two weighted paths $U(i)M(j)U|i = j$ and $U(i)M(j)U||i - j| \leq 1$. $U(i)M(j)U|i = j$ means users rating the exact same scores on the same movies, while $U(i)M(j)U||i - j| \leq 1$ means users rating close scores. Similarly, we design two corresponding weighted paths for UMDMU, UMAMU, and UMTMU. Based on the similarity generated by these meta paths, we employ SemRec$_{Sgl}$ to make recommendations. We compare the performances of SemRec$_{Sgl}$ with different paths and record the results in Fig. 5.6.

The experimental results on all four paths clearly show that SemRec with weighted meta paths (e.g., $U(i)M(j)U|i = j$ and $U(i)M(j)U||i - j| \leq 1$) significantly outperform SemRec with unweighted meta paths (e.g., UMU). Let us take the UMU path as an example to analyze the reasons. Failing to distinguish the different rating scores of users on the same movies, UMU cannot accurately reveal user similarity, so it has bad performances. The path $U(i)M(j)U|i = j$ and $U(i)M(j)U||i - j| \leq 1$ not only considers the differences of rating scores but also keeps dense relations, so they can achieve better performances than UMU. Compared to $U(i)M(j)U|i = j$, the relatively bad performances of $U(i)M(j)U||i - j| \leq 1$ may be attributed to the noise introduced by some improper relation constraints (e.g., U(3)M(4)U, and U(4)M(3)U). The experiments illustrate that the weighted meta paths are really

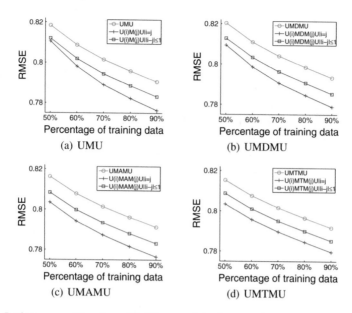

Fig. 5.6 Performances of SemRec with different weighted meta paths

helpful to improve recommendation performances by more accurately revealing object relations.

5.2 Recommendation Based on Matrix Factorization

5.2.1 Overview

With the increasing popularity of social media, there is a surge of social recommendation techniques [4, 12] in recent years, which leverage rich social relations among users, such as friendships in Facebook, following relations in Twitter. However, the emerging social recommendation usually faces the problem of relation sparsity. On the one hand, dense social relations can improve the recommendation performance. However, social relations are very sparse or absent in many real applications. For example, there are no social relations in Amazon, and 80% users in Yelp have less than 3 following relations. On the other hand, users and items in many applications have rich attribute information, which are seldom exploited. These information may be very useful to reveal users' tastes and items' properties. For example, the group attribute of users can reflect their interests, and the type attribute of movies can reveal the content of movies. So it is desirable to effectively integrate all kinds of information for better recommendation performance, including not only feedback and social relations but also attributes of users and items. Some works have began to explore this issue [5, 23, 24], while they did not focus on revealing the importance of these attributes and their effects on recommendation accuracy.

Although integrating more information is promising to achieve better recommendation performance, how to integrate these information still faces two challenges. (1) The information to be integrated has different types. These mixed information types include integer (i.e., rating information), vector (i.e., attribute information), and graph (i.e., social relations). We need to design a unified model to effectively integrate these different types of information. (2) A unified and flexible method is desirable to integrate all or some of these information. In order to intensively study the impacts of different information, the designed method should flexibly integrate different granularities of information and uniformly utilize different types of information.

As mentioned above, we can organize objects and relations in recommended system as a heterogeneous information network which contains different types of nodes or links. In order to utilize these heterogeneous information, we introduce meta path-based similarity measure to evaluate the similarity betweeof users and items. Based on matrix factorization, a dual regularization framework SimMF is proposed to integrate heterogeneous information through adopting similarity information of users and items as regularization on latent factors of users and items. Moreover, in SimMF, two different regularization models, average-based regularization and individual-based regularization, can flexibly confine regularization on users or items.

5.2.2 The SimMF Method

In this section, we will introduce the **SimMF** method, which utilizes **matrix factorization** framework to incorporate **similarity** information. We firstly introduce the rich similarity generation with HIN. And then, we review the basic low-rank matrix factorization framework and introduce the improved model through constraining similarity regularization on users and items, respectively. Finally, we show the unified model through applying similarity regularization on users and items simultaneously.

5.2.2.1 Similarity Generation

Two objects in a heterogeneous network can be connected via different paths, which can be called meta path [19]. A meta path P is a path defined on a schema $S = (A, R)$ and is denoted in the form of $A_1 \xrightarrow{R_1} A_2 \xrightarrow{R_2} \cdots \xrightarrow{R_l} A_{l+1}$ (abbreviated as $A_1 A_2 \cdots A_{l+1}$), which defines a composite relation $R = R_1 \circ R_2 \circ \cdots \circ R_l$ between type A_1 and A_{l+1}, where $\circ$ denotes the composition operator on relations. Since different meta paths have different semantics, objects connecting by different meta paths have different similarity. So we can evaluate the similarity of users (or movies) based on different meta paths. For example, for users, we can consider meta paths

UU, UGU, UMU, and so on. Similarly, meaningful meta paths connecting movies include MAM and MDM.

There are several path-based similarity measures to evaluate the similarity of objects in HIN [6, 16, 19]. Considering semantics in meta paths, Sun et al. [19] proposed PathSim to measure the similarity of same-type objects based on symmetric paths. Lao and Cohen [6] proposed a Path Constrained Random Walk (PCRW) model to measure the entity proximity in a labeled directed graph constructed by the rich metadata of scientific literature. The HeteSim [16] can measure the relatedness of heterogeneous objects based on an arbitrary meta path. All these similarity measures can be used in the similarity calculation, and their differences can be seen in Ref. [16].

We define $S_{ij}^{(l)}$ to denote the similarity of two objects u_i and u_j under the given meta path P_l. The similarity (S) is determined by the given meta path (P) and the similarity measure (M). That is, $S = P \times M$. We know that the similarity of different paths are different and they are incomparable. So we normalize them with *Sigmoid* function as shown in Eq. 5.14, where $\bar{S}^{(l)}$ means the average of $S_{ij}^{(l)}$ and β is set to 1. The normalization process has the following two advantages. (1) It confines the similarity into [0, 1] without changing their ranking. (2) It can reduce the similarity difference of different paths. In the following section, we directly use the $S_{ij}^{(l)}$ to represent the normalized similarity:

$$S_{ij}^{(l)'} = \frac{1}{1 + e^{-\beta \times (S_{ij}^{(l)} - \bar{S}^{(l)})}} \tag{5.14}$$

Since users (or items) have different similarity under different meta paths, we consider their similarity on all paths through assigning weights on different paths. For users, we define S^U for the similarity matrix of users on all paths and S^I for the similarity matrix of items on all paths. They can be defined as follows, where w_l^U represents the weight of similarity matrix of users under the path P_l and w_l^I represents that of items:

$$\begin{aligned} S^U &= \sum_l w_l^U S^{(l)} \quad \sum_l w_l^U = 1; 0 \le w_l^U \le 1 \\ S^I &= \sum_l w_l^I S^{(l)} \quad \sum_l w_l^I = 1; 0 \le w_l^I \le 1 \end{aligned} \tag{5.15}$$

5.2.2.2 Low-Rank Matrix Factorization

The low-rank matrix factorization has been widely studied in recommended system [18]. Its basic idea is to factorize the user–item rating matrix R into two matrices (U and V) representing users' and items' distributions on latent semantic, respectively. Then, the rating prediction can be made through these two specific matrices. Assuming an $m \times n$ rating matrix R to be m users' ratings on n items, this approach mainly minimizes the objective function L(R, U, V) as follows:

$$\min_{U,V} L(R, U, V) = \frac{1}{2} \sum_{i=1}^{m} \sum_{j=1}^{n} I_{ij} (R_{ij} - U_i V_j^T)^2 + \frac{\lambda_1}{2} \|U\|^2 + \frac{\lambda_2}{2} \|V\|^2, \tag{5.16}$$

where I_{ij} is the indicator function that is equal to 1 if user i rates item j and equal to 0 otherwise. $U \in \mathbb{R}^{m \times d}$ and $V \in \mathbb{R}^{n \times d}$, where d is the dimension of latent factors and $d \ll min(m, n)$. U_i is a row vector derived from the ith row of matrix U, and V_j is a row vector derived from the jth row of matrix V. λ_1 and λ_2 represent the regularization parameters. In summary, the optimization problem minimizes the sum-of-squared-errors objective function with quadratic regularization terms which aim to avoid overfitting. This problem can be effectively solved by a simple stochastic gradient descent technique.

5.2.2.3 Similarity Regularization on Users and Items

As mentioned above, the user-specific factorized matrix describes users' distribution over latent semantic. In this section, we will introduce two different types of similarity regularization (i.e., average-based regularization and individual-based regularization) on users to force the distance between U_p and U_q to be much smaller if user p is highly similar to user q.

Average-based Regularization Intuitively, we have similar behavior model with people who are similar with us. That is, the latent factor of a user is similar to the latent factor of people who are the most similar to the user. Based on this assumption, we add user's similarity regularization to the basic low-rank matrix factorization framework.

$$\min_{U,V} \mathrm{L}(R, U, V) = \frac{1}{2} \sum_{i=1}^{m} \sum_{j=1}^{n} I_{ij} (R_{ij} - U_i V_j^T)^2 + \frac{\alpha}{2} \sum_{i=1}^{m} \| U_i - \frac{\sum_{f \in \mathrm{T}_u^+(i)} S_{if}^{\mathrm{U}} U_f}{\sum_{f \in \mathrm{T}_u^+(i)} S_{if}^{\mathrm{U}}} \|^2$$

$$+ \frac{\lambda_1}{2} \| U \|^2 + \frac{\lambda_2}{2} \| V \|^2 \tag{5.17}$$

where $\mathrm{T}_u^+(i)$ is the set of users who are in the top k similarity list of user i and S_{if}^{U} is the element located on the ith row and the fth column of user similarity matrix S^{U}. The average-based regularization confines that the latent factor of a user is close to the average of the latent factor of the top k similar people to the user. The analogous regularization has been used in social recommendation [13], while it just enforces constraints on friends of users. Here, the average-based regularization not only extends to the top k similarity list of users but also considers the similarity values as the weights. The parameter k can be set to trade-off accuracy and computation cost. Large k usually means high accuracy but low efficiency. In our experiments, k is set to 5% of the vector dimension. A local minimum of the objective function given by Eq. 5.17 can be solved by performing gradient descent in feature vectors U_i and V_j, which is shown in Eqs. 5.18 and 5.19. Here, $\mathrm{T}_u^-(i)$ represents the set of users whose top k similarity list contains user i.

$$\frac{\partial L}{\partial U_i} = \sum_{j=1}^{n} I_{ij}(U_i V_j^T - R_{ij})V_j + \alpha(U_i - \frac{\sum_{f \in T_u^+(i)}(S_{if}^U U_f)}{\sum_{f \in T_u^+(i)} S_{if}^U})$$

$$+ \alpha \sum_{g \in T_u^-(i)} \frac{-S_{ig}^U(U_g - \frac{\sum_{f \in T_u^+(g)}(S_{gf}^U U_f)}{\sum_{f \in T_u^+(g)} S_{gf}^U})}{\sum_{f \in T_u^+(g)} S_{gf}^U} + \lambda_1 U_i, \tag{5.18}$$

$$\frac{\partial L}{\partial V_j} = \sum_{i=1}^{m} I_{ij}(U_i V_j^T - R_{ij})U_i + \lambda_2 V_j. \tag{5.19}$$

Individual-based Regularization The above average-based regularization constrains user's taste with the average taste of people who are the most similar users. However, it may be ineffective for users whose similar users have diverse tastes. In order to avoid this disadvantage, we employ individual-based regularization on users as follows:

$$\min_{U,V} L(R, U, V) = \frac{1}{2} \sum_{i=1}^{m} \sum_{j=1}^{n} I_{ij}(R_{ij} - U_i V_j^T)^2 + \frac{\alpha}{2} \sum_{i=1}^{m} \sum_{j=1}^{m} S_{ij}^U \|U_i - U_j\|^2$$

$$+ \frac{\lambda_1}{2} \|U\|^2 + \frac{\lambda_2}{2} \|V\|^2. \tag{5.20}$$

In essential, the individual-based regularization enforces a large S_{ij}^U to have a small distance between U_i and U_j. That is, similar users have smaller distance on latent factors. With the same optimization technique, a local minimum of Eq. 5.20 can also be found by performing gradient descent in U_i and V_j.

$$\frac{\partial L}{\partial U_i} = \sum_{j=1}^{n} I_{ij}(U_i V_j^T - R_{ij})V_j + \alpha \sum_{j=1}^{m}(S_{ij}^U + S_{ji}^U)(U_i - U_j) + \lambda_1 U_i, \tag{5.21}$$

$$\frac{\partial L}{\partial V_j} = \sum_{i=1}^{m} I_{ij}(U_i V_j^T - R_{ij})U_i + \lambda_2 V_j. \tag{5.22}$$

Similarity Regularization on Items For simplicity, we define the notation Reg_y^x to represent the average-based or individual-based regularization term on users or items, where $x \in \{U, I\}$ means Users or Items and $y \in \{ave, ind\}$ means *ave*rage-based or *ind*ividual-based regularization. That is, for similarity regularization on users, we have

$$Reg_{ave}^{U} = \sum_{i=1}^{m} \| U_i - \frac{\sum_{f \in T_u^+(i)} S_{if}^{U} U_f}{\sum_{f \in T_u^+(i)} S_{if}^{U}} \|^2, \tag{5.23}$$

$$Reg_{ind}^{U} = \sum_{i=1}^{m} \sum_{j=1}^{m} S_{ij}^{U} \| U_i - U_j \|^2. \tag{5.24}$$

Similar to the regularization on users, we can also define these two different types of regularization on items as follows:

$$Reg_{ave}^{I} = \sum_{j=1}^{n} \| V_j - \frac{\sum_{f \in T_i^+(j)} S_{jf}^{I} V_f}{\sum_{f \in T_i^+(j)} S_{jf}^{I}} \|^2, \tag{5.25}$$

$$Reg_{ind}^{I} = \sum_{i=1}^{n} \sum_{j=1}^{n} S_{ij}^{I} \| V_i - V_j \|^2, \tag{5.26}$$

where $T_i^+(j)$ is the set of items who are in the top k similarity list of item j, and S_{jf}^{I} is the element located on the jth row and the fth column of similarity matrix S^I. We can also define the optimization function based on these two regularization terms on items and derive their gradient learning algorithms as above.

5.2.2.4 A Unified Dual Regularization

Now, we consider regularization on users and items simultaneously. The corresponding optimization function is shown as follows:

$$\min_{U,V} \mathrm{L}(R, U, V) = \frac{1}{2} \sum_{i=1}^{m} \sum_{j=1}^{n} I_{ij} (R_{ij} - U_i V_j^T)^2 + \frac{\alpha}{2} Reg_y^{U} + \frac{\beta}{2} Reg_y^{I}$$

$$+ \frac{\lambda_1}{2} \| U \|^2 + \frac{\lambda_2}{2} \| V \|^2, \tag{5.27}$$

where α and β control the effect of user and item regularization, respectively. For $y \in \{ave, ind\}$, there are four regularization models. Similarly, we can use the gradient descent method to solve this optimization problem. The whole algorithm framework is shown in Algorithm 2.

5.2.3 Experiments

In this section, we will verify the superiority of our model by conducting a series of experiments compared to the state-of-the-art recommendation methods.

Algorithm 2 Algorithm Framework of SimMF

Require:
 G: heterogeneous information network
 P_U, P_I: sets of meta paths related to users and items
 η: learning rate for gradient descent
 $\alpha, \beta, \lambda_1, \lambda_2$: controlling parameters defined above
 ε: convergence tolerance
Ensure:
 U, V: the latent factor of users and items

1: Calculate similarity matrix of user S_U based on P_U, G
2: Calculate similarity matrix of item S_I based on P_I, G
3: Initialize U, V
4: **repeat**
5:   $U_{old} := U, V_{old} := V$
6:   Calculate $\frac{\partial \mathrm{L}}{\partial U}, \frac{\partial \mathrm{L}}{\partial V}$
7:   Update $U := U - \eta * \frac{\partial \mathrm{L}}{\partial U}$
8:   Update $V := V - \eta * \frac{\partial \mathrm{L}}{\partial V}$
9: **until** $\|U - U_{old}\|^2 + \|V - V_{old}\|^2 < \varepsilon$

5.2.3.1   Experiment Settings

In experiments, we employs two real datasets from two various domains. Douban Movie[3] is from the movie domain. Stemming from the business domain, the widely used Yelp challenge dataset[4] [23, 24] records users' ratings on local business and also contains social relations and attribute information of business (e.g., cities and categories). In addition, we use Mean Absolute Error (MAE) and Root Mean Square Error (RMSE) to evaluate the performance of different methods.

In this section, we compare SimMF with six representative methods. There are different variations for SimMF. We use SimMF-U(y)I(y) to represent SimMF with regularization on users and items, where $y \in \{a, i\}$, and it represents the average-based or individual-based regularization. Similarly, SimMF-U(y) (SimMF-I(y)) means SimMF with regularization only on users (items). There are six baseline methods, including four types. There are two basic methods (i.e., UserMean and ItemMean), a collaborative filtering with low-rank matrix factorization (i.e., PMF), a social recommendation method (i.e., SoMF), and two HIN-based methods (i.e., HeteMF and HeteCF). These baselines are summarized as follows.

- **UserMean**. This method uses the mean value of every user to predict the missing values.
- **ItemMean**. This method utilizes the mean value of every item to predict the missing values.

---

[3]http://movie.douban.com/.

[4]http://www.yelp.com/dataset_challenge/.

- **PMF**. This method is a typical matrix factorization method proposed by Salakhut-dinov and Minh [15]. And in fact, it is equivalent to basic low-rank matrix factor-ization in the previous section.
- **SoMF**. This is the matrix factorization-based recommendation method with social average-based regularization proposed by Ma et al. [13].
- **HeteMF**. This is the matrix factorization-based recommendation framework com-bining user ratings and various entity similarity matrices proposed by Yu et al. [22].
- **HeteCF**. This is the social collaborative filtering algorithm using heterogeneous relations [9].

We employ HeteSim [16] to evaluate the similarity of objects. For the Douban Movie dataset, we use 7 meaningful meta paths for user whose length is smaller than 4 (i.e., UU, UGU, ULU, UMU, UMDMU, UMTMU, UMAMU) and 5 meaningful meta paths for movie whose length is smaller than 3 (i.e., MTM, MDM, MAM, MUM, MUUM). For the Yelp dataset, we use 4 meta paths for user (i.e., UU, UBU, UBCBU, UBLBU) and 4 meta paths for business (i.e., BUB, BCB, BLB, BUUB). Similarly, we utilize 5 meta paths for user (i.e., UGU, UAU, UOU, UMU, UMTMU) and 2 meta paths for movie (i.e., MTM, MUM) for the MovieLens dataset. And for the Douban Book dataset, we utilize 7 meta paths for user (i.e., UU, UGU, ULU, UBU, UBABU, UBPBU, UBYBU) and 5 meta paths for book (i.e., BAB, BPB, BYB, BUB, BUUB). These similarity data are fairly used for HeteCF and SimMF. HeteMF uses similarity data of users, since the model only considers the similarity relationships between items.

5.2.3.2   Effectiveness Experiments

This section will validate the effectiveness of SimMF through comparing its different variations to baselines. Here, we run four versions of SimMF-U(y)I(y) ($y \in \{a, i\}$) and record the worst (denoted as SimMF-max in Tables 5.5 and 5.6), the best (denoted as SimMF-min), and the average (denoted as SimMF-mean) performance of these four versions. The α and β are set to 100 and 10, respectively, for Douban Movie dataset, as suggested in the following parameter experiment. For other datasets, α and β are set to the optimal values according to related parameter experiments. For all the experiments in this chapter, the values of λ_1 and λ_2 are set to a trivial value 0.001 and the length of latent feature vectors U_i and V_j are set to 10. The parameters of other methods are set to the optimal values obtained in parameter experiments.

For these datasets, we use different ratios (80%, 60%, 40%, 20%) of data as training set. For example, the training data 80% means that we select 80% of the ratings from user–item rating matrix as the training data to predict the remaining 20% of ratings. The random selection was carried out 10 times independently in all the experiments. We report average results on Douban Movie and Yelp datasets in Tables 5.5 and 5.6, respectively, and record the improvement ratio of all meth-ods compared to the PMF. In addition, we also report the average running time of these methods with the 80% training ratio in the last line of above tables. For those

HIN-based methods (i.e., HeteCF, HeteMF, and SimMF), we only report the running time of the model learning process, ignoring the running time of similarity computation. Note that we report the mean running time for SimMF, since the four versions of SimMF have the similar computational complexity.

The results are shown in Tables 5.5 and 5.6. In addition, we also conduct the t-test experiments with 95% confidence, which shows that the MAE/RMSE improvement difference is statistically stable and non-contingent. Due to the space limitation, they are omitted in the paper, but the results can be found in [17]. From the experimental comparisons, we can observe the following phenomena.

- SimMF always outperforms the baselines in most conditions, even for the worst performance of SimMF (i.e., SimMF-max). It validates that more attribute information from users and items exploited in SimMF is really helpful to improve the recommendation performance. In addition, the model integrating more information usually has better performances. That is, the reason why other matrix factorization models integrating heterogeneous information usually have better performance than the basic matrix factorization model PMF.
- Although HeteMF and HeteCF also utilize the attribute information from users and items, they have worse performance than SimMF, which implies the proposed SimMF has better mechanism to integrate heterogeneous information. We know that HeteMF only integrates attribute information of items, while the same parameter for similarity regularization terms of users and items may cause the bad performance of HeteCF.
- When considering different training data ratios, we can find that the superiority of SimMF is more significant for less training data. It indicates that SimMF can effectively alleviate data sparsity problem. We think the reason lies in that, through exploiting different meta paths, we can make full use of rich attribute information of users and items to reflect the similarity of users and items from different aspects. The integration of similarities can comprehensively reveal the similarity of users and items, which compensates for shortage of training data.

Observing the running time of different methods in the last row of Tables 5.5 and 5.6, we can find that the running time becomes longer as the models become more complex. That is, HIN-based methods (i.e., HeteMF, HeteCF, and SimMF) have longer running time than the other methods, since they have more parameters to be learned. However, SimMF is still faster than the other two HIN-based methods because SimMF does not need to learn the weights of meta paths.

5.2.3.3 Impact of Different Regularizations

Experiments in this section will validate the effect of different regularization models on users and items. Ma et al. [13] have explored the effect of average-based and individual-based regularization on social relations of users. However, in this chapter, we not only explore the effect on more complex relations, but also consider the effect on both users and items.

Table 5.5 Performance comparisons on Douban Movie (the baseline of improved performance is PMF)

Training	Metrics	UserMean	ItemMean	PMF	SoMF	HeteMF	HeteCF	SimMF-mean	SimMF-max	SimMF-min
80%	MAE	0.6958	0.6476	0.6325	0.6073	0.6221	0.6273	0.5974	0.6026	**0.5926**
	Improve	−10.01%	−2.83%		3.99%	1.64%	0.82%	5.55%	4.73%	6.31%
	RMSE	0.8846	0.8537	0.8815	0.8283	0.8609	0.8664	0.7729	0.7809	**0.7656**
	Improve	−0.35%	3.15%		6.03%	2.34%	1.71%	12.32%	11.41%	13.14%
60%	MAE	0.6986	0.6557	0.6591	0.6219	0.6490	0.6509	0.6060	0.6110	**0.6008**
	Improve	−6.00%	0.35%		5.63%	1.53%	1.24%	8.06%	7.30%	8.85%
	RMSE	0.8925	0.8748	0.9281	0.8584	0.9100	0.9118	0.7852	0.7927	**0.7772**
	Improve	3.84%	5.75%		7.51%	1.95%	1.76%	15.40%	14.59%	16.26%
40%	MAE	0.7052	0.6733	0.7092	0.6457	0.6933	0.7029	0.6186	0.6237	**0.6134**
	Improve	0.57%	5.07%		8.96%	2.24%	0.89%	12.77%	12.06%	13.51%
	RMSE	0.9085	0.9139	1.0107	0.9034	0.9842	0.9941	0.8023	0.8093	**0.7952**
	Improve	10.11%	9.57%		10.62%	2.62%	1.64%	20.62%	19.93%	21.32%
20%	MAE	0.7227	0.7124	0.8367	0.6973	0.8235	0.8302	0.6461	0.6509	**0.6417**
	Improve	13.63%	14.85%		16.66%	1.58%	0.78%	22.78%	22.21%	23.31%
	RMSE	0.9502	1.0006	1.2060	1.0037	1.1838	1.1963	0.8388	0.8446	**0.8335**
	Improve	21.21%	17.03%		16.78%	1.84%	0.80%	30.45%	29.97%	30.89%
Running time(s)		0.5157	0.5242	1096	1385	4529	7342	3168		

Table 5.6 Performance comparisons on Yelp (the baseline of improved performance is PMF)

Training	Metrics	UserMean	ItemMean	PMF	SoMF	HeteMF	HeteCF	SimMF-mean	SimMF-max	SimMF-min
80%	MAE	0.9664	0.8952	1.2201	0.8789	0.9307	1.2117	0.8292	0.8503	**0.8059**
	Improve	20.79%	26.63%		27.96%	23.72%	0.69%	32.04%	30.31%	33.95%
	RMSE	1.3443	1.2327	1.6479	1.1912	1.2773	1.6249	1.0577	1.0708	**1.0465**
	Improve	18.42%	25.20%		27.71%	22.49%	1.40%	35.82%	35.02%	36.49%
60%	MAE	0.9803	0.9247	1.3835	0.9156	0.9708	1.3510	0.8366	0.8615	**0.8109**
	Improve	29.14%	33.16%		33.82%	29.83%	2.35%	39.53%	37.73%	41.39%
	RMSE	1.3556	1.2893	1.8438	1.2591	1.3352	1.7940	1.0684	1.0842	**1.0532**
	Improve	26.48%	30.07%		31.71%	27.58%	2.70%	42.05%	41.20%	42.88%
40%	MAE	1.0219	0.9819	1.7081	0.9790	1.0409	1.6360	0.8509	0.8810	**0.8186**
	Improve	40.17%	42.52%		42.68%	39.06%	4.22%	50.18%	48.42%	52.18%
	RMSE	1.4241	1.3873	2.2123	1.3682	1.4343	2.1116	1.0863	1.1031	**1.0686**
	Improve	35.63%	37.29%		38.15%	35.17%	4.55%	50.90%	50.12%	51.70%
20%	MAE	1.1344	1.1202	2.6935	1.1252	1.8429	2.5782	0.8687	0.9047	**0.8290**
	Improve	57.88%	58.41%		58.23%	31.58%	4.28%	67.75%	66.41%	69.22%
	RMSE	1.5958	1.5981	3.2512	1.5907	2.3357	3.0807	1.1307	1.1733	**1.0944**
	Improve	50.92%	50.85%		51.07%	28.16%	5.24%	65.22%	63.91%	66.34%
Running time(s)		0.0646	0.0642	100	137	1963	2378	1414		

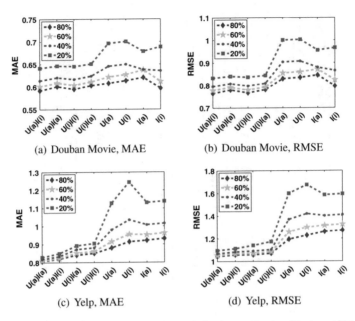

Fig. 5.7 Performance of SimMF with different regularizations on Douban Movie and Yelp datasets

We employ four variations of SimMF with average-based and individual-based regularization on users and items (i.e., SimMF with U(a)I(i), U(a)I(a), U(i)I(i), and U(i)I(a)) and four variations of SimMF with average-based or individual-based regularization on users or items (i.e., SimMF with U(a), U(i), I(a), and I(i)). The same parameters are set with above experiments, and the average results are shown in Fig. 5.7. We can find that SimMF, integrating similarity information on both users and items, always has better performance than the one only integrating similarity information on users or items. Again, we can observe the difference is far more pronounced when the fraction of training set is low; e.g., at 20%, SimMF-U(i) and SimMF-U(a) perform very bad. Moreover, we can also observe an interesting phenomena: Regularization models have different effects on users and items. SimMF-U(a) has better performance than SimMF-U(i) on both datasets, which indicates average-based regularization may be more suitable for users. However, it is not the case for items. SimMF-I(i) performs better than SimMF-I(a) on Douban Movie, while SimMF-I(a) outperforms SimMF-I(i) on Yelp. As a result, SimMF-U(a)I(i) has the best performance on Douban Movie, while SimMF-U(a)I(a) is the best one on Yelp. Although it is hard to draw general conclusions, the above study indicates that different regularization model may significantly affect performance of matrix factorization methods. In summary, we need to find the optimal regularization model according to data properties in real applications.

5.2.3.4 Impact of Different Meta Paths

In this section, we study the impact of different meta paths. Due to similar analysis, we only show results on Douban Movie dataset. As illustrated above, we employ 7 meta paths on users and 5 meta paths on movies. We will observe performance of SimMF with similarity matrix generated by one single meta path. Under the same parameters with above experiments, we run SimMF-U(a) with similarity matrix generated by each meta path on users. Similarly, we also run SimMF-I(i) with similarity matrix generated by each meta path on movies.

The experiment results on Douban Movie dataset are shown in Fig. 5.8. We can observe different impacts of meta paths on users and movies. The SimMF-U(a) with different meta paths (see Fig. 5.8a, b) on users all have close performance. Moreover, SimMF-U(a) with MUM has slightly better performance and SimMF-U(a) with UU has worse performance. However, it is not the case for meta paths on items. The SimMF-I(i) with different meta paths on items (see Fig. 5.8c, d) have totally different performance. We can find that SimMF-I(i) with MDM has the worst performance, even worse than PMF in some conditions, while SimMF-I(i) with MTM and MUM achieve much better performance on both criteria. We think there are two reasons: (1) Note that the performance of SimMF is much affected by the density of relations. The density of relations on MT and MU is much higher than that on MD and MA. The dense relations are helpful to generate good similarity of items. The similar

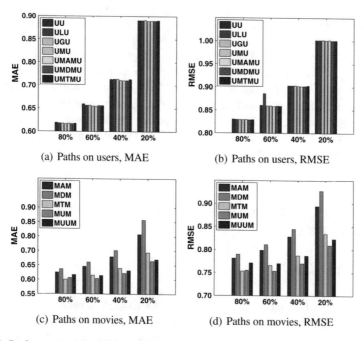

(a) Paths on users, MAE   (b) Paths on users, RMSE

(c) Paths on movies, MAE   (d) Paths on movies, RMSE

Fig. 5.8 Performance of SimMF with different meta paths on Douban Movie dataset

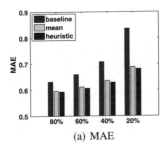
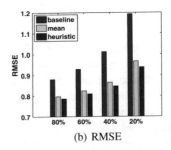

(a) MAE   (b) RMSE

Fig. 5.9 Performance of SimMF on MAE and RMSE with different weights setting methods

phenomena have been widely observed in social recommendation [10, 13]. (2) The meaningful meta paths are helpful to reveal the similarity of objects. MTM means movies with same type, and MUM means movies seen by same users. These two paths are highly correlated as both reveal properties of the movies. These two reasons can also explain the slightly worse performance of the meaningful but sparse UU meta path as compared to other meta paths of users. The experiments imply that we only need to use one single dense and meaningful meta path to generate similarity information, which also can obtain good enough performance.

We further design an experiment to illustrate different importance of meta paths. Concretely, we observe the performance of above SimMF-I(i) with different weight combination methods on 5 meta paths. Except mean weight and random weight on 5 paths, we design a heuristic weight method, i.e., setting the weights according to the performance of these paths. That is, paths with good performance have higher weights. Assume the MAE performance value of a path (P_l) is P_l, and the max MAE value is P_{max}. Then, the difference is $d_l = e^{P_{max}-P_l}$. And thus, the weight of the path is $w_l^{\mathsf{T}} = \frac{d_l}{\sum_l d_l}$. The experiment also includes PMF as the baseline. The results are shown in Fig. 5.9. It is obvious that SimMF-I(i) with the heuristic weight method has the best performance, which further validates that the meaningful and dense meta paths are more important. The more detailed method description and experiment validation can be seen in [17].

5.3 Social Recommendation with Heterogeneous Information

5.3.1 Overview

With the boom of social media, social recommendation has become a hot research topic, which utilizes the social relations among users for better recommendation. Some researchers utilized trust information among users [10, 11], and some began to use friend relationship among users [13, 21] or other types of information

[1, 2]. Most of these social recommendation methods employ social regularization to confine similar users under the low-rank matrix factorization framework. Specifically, we can obtain the similarity of users from their social relations, and then, the social regularization, as a constraint term, confines the latent factors of similar users to be closer. It is reasonable, since similar users should have similar latent features.

However, the widely used social regularization in social recommendation has several shortcomings. (1) The similarity information of users is only generated from social relations of users. But we can obtain users' similarity from many ways in real applications, such as users' contents. (2) The social regularization only has constraint on users. In fact, we can also obtain the similarity of items and impose constraint on the latent factors of items. (3) The social regularization may be ineffective for dissimilar users, which may lead to dissimilar users having similar factors. The analysis and experiments in the next section validate this point.

In order to address the limitation of traditional social recommendation, we propose a dual similarity regularization-based recommendation method (called DSR). Inspired by the success of Heterogeneous Information Network (HIN) in many applications, we organize a recommended system as an HIN, which can integrate all kinds of information, including interactions between users and items, social relations among users, and attribute information of users and items. Based on the HIN, we can generate rich similarity information on both users and items by setting proper meta paths. Furthermore, we propose a new similarity regularization which can impose the constraint on users and items with high and low similarity. With the similarity regularization, DSR adopts a new optimization objective to integrate those similarity information of users and items. Then, we derive its solution to learn the weights of different similarities.

5.3.2 The DSR Method

In this section, we propose the dual similarity regularization-based matrix factorization method **DSR** and infer its learning algorithm.

5.3.2.1 Limitations of Social Recommendation

Recently, with the increasing popularity of social media, there is a surge of social recommendations which leverage rich social relations among users to improve recommendation performance. Ma et al. [13] first proposed the social regularization to extend low-rank matrix factorization, and then, it is widely used in a lot of work [9, 22]. A basic social recommendation method is illustrated as follows:

$$\min_{U,V} \mathcal{J} = \frac{1}{2} \sum_{i=1}^{m} \sum_{j=1}^{n} I_{ij} (R_{ij} - U_i V_j^T)^2 + \frac{\alpha}{2} \sum_{i=1}^{m} \sum_{j=1}^{m} S_U(i, j) \|U_i - U_j\|^2$$
$$+ \frac{\lambda_1}{2} (\|U\|^2 + \|V\|^2), \tag{5.28}$$

where $m \times n$ rating matrix R depicts users' ratings on n items and R_{ij} is the score user i gives to item j. I_{ij} is an indicator function which equals to 1 if user i rated item j and equals to 0 otherwise. $U \in \mathbb{R}^{m \times d}$ and $V \in \mathbb{R}^{n \times d}$, where $d \ll min(m, n)$ is the dimension number of latent factor. U_i is the latent vector of user i derived from the ith row of matrix U, while V_j is the latent vector of item j derived from the jth row of V. S_U is the similarity matrix of users, and $S_U(i, j)$ denotes the similarity of user i and user j. $\| \cdot \|^2$ is the Frobenius norm. Particularly, the second term is the social regularization which is defined as follows:

$$SocReg = \frac{1}{2} \sum_{i=1}^{m} \sum_{j=1}^{m} S_U(i, j) \|U_i - U_j\|^2. \tag{5.29}$$

As a constraint term in Eq. 5.28, $SocReg$ forces the latent factors of two users to be close when they are very similar. However, it may have two drawbacks.

- The similarity information may be simple. In social recommendation, the similarity information of users is usually generated from rating information or social relations, and only one type of similarity information is employed. However, in many applications, we can obtain much more rich similarity information of users and items from various ways, such as rich attribute information and interactions. We need to make full use of these similarity information of users and items for recommendation.
- The constraint term may not work well when two users are not very similar. The minimization of optimization objective should force the latent factors of two users with high similarity to be close. However, when two users are not similar (i.e., $S_U(i, j)$ is small), $SocReg$ may still force the latent factors of these two users to be close. However, these two users should be dissimilar which means their latent factors should have large distance.

In order to uncover the limitations of social regularization, we apply the model detailed in Eq. 5.28 to conduct four experiments each with different levels of similarity information (*None, Low, High, All*). *None* denotes that we utilize no similarity information in the model (i.e., $\alpha = 0$ in the model), *Low* denotes that we utilize bottom 20% users' similarity information generated in the model, *High* is that of top 20%, and *All* denotes we utilize all users' similarity information. The Douban dataset is employed in the experiments, and we report MAE and RMSE in Fig. 5.10. The results of *Low, High,* and *All* are better than that of *None*, which implies social regularization really works in the model. However, in terms of performance improvement compared to *None, Low* does not improve as much as *High* and *All*

Fig. 5.10 Limitations of social regularization

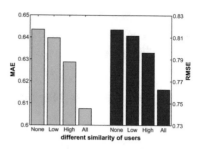

do. The above analysis reveals that the social regularization may not work well in recommender models when users are with low similarity.

5.3.2.2 Rich Similarity Generation

Traditional social recommendations only consider the constraint of users with their social relations. However, rich similarity information on users and items can be generated in a heterogeneous information network. Two types of objects in an HIN can be connected via various meta paths [19], which is a composite relation connecting these two types of objects. Therefore, we can evaluate the similarity of users (or movies) based on different meta paths. For example, for users, we can consider UU, UGU, UMU, etc. Similarly, meaningful meta paths connecting movies include MAM and MDM.

Several path-based similarity measures have been proposed to evaluate the similarity of objects under given meta path in HIN [16, 19]. We assume that $S_U^{(p)}$ denotes similarity matrix of users under meta path $P_U^{(p)}$ connecting users, and $S_U^{(p)}(i, j)$ denotes the similarity of users i and j under the path $P_U^{(p)}$. Similarly, $S_I^{(q)}$ denotes similarity matrix of items under the path $P_I^{(q)}$ connecting items, and $S_I^{(q)}(i, j)$ denotes the similarity of items i and j. Since users (or items) have different similarities under various meta paths, we combine their similarities on all paths through assigning weights on these paths. For users and items, we define S_U and S_I to represent the similarity matrix of users and items on all meta paths, respectively.

$$S_U = \sum_{p=1}^{|P_U|} w_U^{(p)} S_U^{(p)}, \tag{5.30}$$

$$S_I = \sum_{q=1}^{|P_I|} w_I^{(q)} S_I^{(q)}, \tag{5.31}$$

where $w_U^{(p)}$ denotes the weight of meta path $P_U^{(p)}$ among all meta paths P_U connecting users, and $w_I^{(q)}$ denotes the weight of meta path $P_I^{(q)}$ among all meta paths P_I connecting items.

5.3.2.3   Similarity Regularization

Due to the limitations of widely used social regularization, we design a new similarity regularization to constraining users and items simultaneously with much available similarity information of users and items. The basic idea of similarity regularization is that the distance of latent factors of two users (or items) should be negatively correlated to their similarity, which means two similar users (or items) should have a short distance while two dissimilar ones should have a long distance with their latent factors. We note that the Gaussian function meets above requirement. Moreover, the range of Gaussian function is [0,1], same with the range of similarity function. Following this idea, we design a similarity regularization on users as follows:

$$SimReg^U = \frac{1}{8} \sum_{i=1}^{m} \sum_{j=1}^{m} (S_U(i, j) - e^{-\gamma \|U_i - U_j\|^2})^2, \qquad (5.32)$$

where γ controls the radial intensity of Gaussian function and the coefficient $\frac{1}{8}$ is convenient for deriving the learning algorithm. This similarity regularization can enforce constraint on both similar and dissimilar users. In addition, the similarity matrix S_U can be generated from social relations or the above HIN. Similarly, we can also design the similarity regularization on items as follows:

$$SimReg^I = \frac{1}{8} \sum_{i=1}^{n} \sum_{j=1}^{n} (S_I(i, j) - e^{-\gamma \|V_i - V_j\|^2})^2, \qquad (5.33)$$

The Proposed DSR Model We propose the **D**ual **S**imilarity regularization for **R**ecommendation (called DSR) through adding the similarity regularization on users and items into low-rank matrix factorization framework. Specifically, the optimization model is proposed as follows:

$$\min_{U,V,w_U,w_I} \mathcal{J} = \frac{1}{2} \sum_{i=1}^{m} \sum_{j=1}^{n} I_{ij}(R_{ij} - U_i V_j^T)^2$$

$$+ \frac{\lambda_1}{2}(\|U\|^2 + \|V\|^2) + \frac{\lambda_2}{2}(\|w_U\|^2 + \|w_I\|^2)$$

$$+ \alpha SimReg^U + \beta SimReg^I \qquad (5.34)$$

$$s.t. \quad \sum_{p=1}^{|P_U|} w_U^{(p)} = 1, w_U^{(p)} \geq 0$$

$$\sum_{q=1}^{|P_I|} w_I^{(q)} = 1, w_I^{(q)} \geq 0,$$

where α and β control the ratio of similarity regularization term on users and items, respectively.

5.3.2.4 The Learning Algorithm

The learning algorithm of DSR can be divided into two steps. (1) Optimize the latent factor matrices of users and items (i.e., U, V) with the fixed weight vectors $\boldsymbol{w}_U = [w_U^{(1)}, w_U^{(2)}, \cdots, w_U^{(|P_U|)}]^T$ and $\boldsymbol{w}_I = [w_I^{(1)}, w_I^{(2)}, \cdots, w_I^{(|P_I|)}]^T$. (2) Optimize the weight vectors $\boldsymbol{w}_U$ and $\boldsymbol{w}_I$ with the fixed latent factor matrices U and V. Through iteratively optimizing these two steps, we can obtain the optimal U, V, $\boldsymbol{w}_U$, and $\boldsymbol{w}_I$.

Optimize U and V With the fixed $\boldsymbol{w}_U$ and $\boldsymbol{w}_I$, we can optimize U and V by performing stochastic gradient descent.

$$\frac{\partial \mathcal{J}}{\partial U_i} = \sum_{j=1}^{n} I_{ij}(U_i V_j^T - R_{ij})V_j \tag{5.35}$$

$$+\alpha \sum_{j=1}^{m} \gamma[(S_U(i, j) - e^{-\gamma\|U_i - U_j\|^2})e^{-\gamma\|U_i - U_j\|^2}(U_i - U_j)]$$

$$+\lambda_1 U_i,$$

$$\frac{\partial \mathcal{J}}{\partial V_j} = \sum_{i=1}^{m} I_{ij}(U_i V_j^T - R_{ij})U_i \tag{5.36}$$

$$+\beta \sum_{i=1}^{n} \gamma[(S_I(i, j) - e^{-\gamma\|V_i - V_j\|^2})e^{-\gamma\|V_i - V_j\|^2}(V_i - V_j)]$$

$$+\lambda_1 V_j,$$

Optimize $\boldsymbol{w}_U$ and $\boldsymbol{w}_I$ With the fixed U and V, the minimization of $\mathcal{J}$ with respect to $\boldsymbol{w}_U$ and $\boldsymbol{w}_I$ is a well-studied quadratic optimization problem with nonnegative bound. We can use the standard trust region reflective algorithm to update $\boldsymbol{w}_U$ and $\boldsymbol{w}_I$ at each iteration. We can simplify the optimization function of $\boldsymbol{w}_U$ as the following standard quadratic formula:

$$\min_{\mathbf{w}_U} \frac{1}{2} \mathbf{w}_U^T H_U \mathbf{w}_U + f_U^T \mathbf{w}_U \qquad (5.37)$$

$$s.t. \sum_{p=1}^{|P_U|} \mathbf{w}_U^{(p)} = 1, \mathbf{w}_U^{(p)} \geq 0.$$

Here, H_U is a $|P_U| \times |P_U|$ symmetric matrix as follows:

$$H_U(i, j) = \begin{cases} \frac{\alpha}{4}(\sum \sum S_U^{(i)} \odot S_U^{(j)}) & i \neq j, \ 1 \leq i, j \leq |P_U| \\ \frac{\alpha}{4}(\sum \sum S_U^{(i)} \odot S_U^{(j)}) + \lambda_2 & i = j, \ 1 \leq i, j \leq |P_U|, \end{cases}$$

$\odot$ denotes the dot product. f_U is a column vector with length $|P_U|$, which is calculated as follows:

$$f_U(p) = -\frac{\alpha}{4} \sum_{i=1}^{m} \sum_{j=1}^{m} S_U^{(p)}(i, j) e^{-\gamma \|U_i - U_j\|^2}.$$

Similarly, we can also infer the optimization function of $\mathbf{w}_I$.

5.3.3 Experiments

In this section, we conduct experiments to validate the effectiveness of DSR and further explore the cold-start problem.

5.3.3.1 Experiment Settings

We use two real datasets: Douban and Yelp in experiments. Note that the Douban dataset has sparse social relationship with dense rating information, while the Yelp dataset has dense social relationships with sparse rating information. We still use Mean Absolute Error (MAE) and Root Mean Square Error (RMSE) to evaluate the performance of rating prediction.

In order to validate the effectiveness of DSR, we compare it with following representative methods. Besides the classical social recommendation method SoMF, the experiments also include two recent HIN-based methods, HeteCF and HeteMF. In addition, in order to validate the effectiveness of similarity regularization, we include the revised version of SoMF with similarity regularization (i.e., SoMF$_{SR}$).

- **UserMean**. It employs a user's mean rating to predict the missing ratings directly.
- **ItemMean**. It employs an item's mean rating to predict the missing ratings directly.
- **PMF** [14]. Salakhutdinov and Minh proposed the basic low-rank matrix factorization method for recommendation.

- **SoMF** [13]. Ma et al. proposed the social recommendation method with social regularization on users.
- **HeteCF** [9]. Luo et al. proposed the social collaborative filtering algorithm using heterogeneous relations.
- **HeteMF** [22]. Yu et al. proposed the HIN-based recommendation method through combining user ratings and items' similarity matrices.
- **SoMF$_{SR}$**. It adapts SoMF through only replacing the social regularization with the similarity regularization $SimReg^U$.

For Douban dataset, we utilize 7 meta paths for user (i.e., UU, UGU, ULU, UMU, UMDMU, UMTMU, and UMAMU) and 5 meta paths for item (i.e., MTM, MDM, MAM, MUM, and MUUM). For Yelp dataset, we utilize 2 meta paths for user (i.e., UB and UU) and 2 meta paths for item (i.e., BC and BL). HeteSim [16] is employed to evaluate the object similarity based on above meta paths. These similarity matrices are fairly utilized for HeteCF, HeteMF, and DSR. We set $\gamma = 1$, $\alpha = 10$, and $\beta = 10$ through parameter experiments on Douban dataset. In the experiments on Yelp dataset, we set the parameters $\gamma = 1$, $\alpha = 10$, and $\beta = 10$. Meanwhile, optimal parameters are set for other models in the experiments.

5.3.3.2 Effectiveness Experiments

For Douban dataset, we use different ratios (80%, 60%, 40%) of data as training sets and the rest of the dataset for testing. Considering the sparse density of Yelp dataset, we use 90%, 80%, and 70% of data as training sets and the rest of the dataset for testing for Yelp dataset. The random selection is carried out 10 times independently, and we report the average results in Table 5.7.

It is clear that three HIN-based methods (DSR, HeteCF, and HeteMF) all achieve significant performance improvements compared to PMF, UserMean, ItemMean, and SoMF. It implies that integrating heterogeneous information is a promising way to improve recommendation performance. Particularly, DSR always has the best performance on all conditions compared to other methods. It indicates that the dual similarity regularization on users and items may be more effective than traditional social regularization. It can be further confirmed by the better performance of SoMF$_{SR}$ over SoMF. Although the superiority of SoMF$_{SR}$ over SoMF is not significant, the improvement is achieved on the very weak social relations in Douban dataset. In addition, we can also find that DSR has better performance improvement for less training data. It reveals that DSR has the potential to alleviate the cold-start problem.

5.3.3.3 Study on Cold-Start Problem

Furthermore, we validate the superiority of DSR on cold-start problem. We run PMF, SoMF, HeteCF, HeteMF, and DSR on Douban dataset with 40% training ratio. We set four levels of users: three types of cold-start users with various numbers

Table 5.7 Effectiveness experimental results on Douban and Yelp (The improvement is based on PMF)

Dataset	Training	Metrics	PMF	UserMean	ItemMean	SoMF	HeteCF	HeteMF	SoMF$_{SR}$	DSR
Douban	80%	MAE	0.6444	0.6954	0.6284	0.6396	0.6101	0.5941	0.6336	**0.5856**
		Improve		−7.92%	2.47%	0.73%	5.32%	7.79%	1.68%	9.12%
		RMSE	0.8151	0.8658	0.7928	0.8098	0.7657	0.7520	0.8000	**0.7379**
		Improve		−6.23%	2.73%	0.64%	6.05%	7.73%	1.85%	9.46%
	60%	MAE	0.6780	0.6967	0.6370	0.6696	0.6317	0.6056	0.6648	**0.5946**
		Improve		−2.76%	6.05%	1.25%	6.84%	10.68%	1.96%	12.31%
		RMSE	0.8569	0.8687	0.8135	0.8445	0.7901	0.7665	0.8358	**0.7483**
		Improve		−1.37%	5.07%	1.45%	7.80%	10.56%	2.46%	12.68%
	40%	MAE	0.7364	0.7009	0.6629	0.7245	0.6762	0.6255	0.7141	**0.6092**
		Improve		4.83%	9.99%	1.63%	8.18%	15.07%	3.03%	17.28%
		RMSE	0.9221	0.8747	0.8747	0.9058	0.8404	0.7891	0.8950	**0.7629**
		Improve		5.14%	5.13%	1.76%	8.86%	14.42%	2.94%	17.27%
Yelp	90%	MAE	0.8475	0.9543	0.8822	0.8460	0.8461	0.8960	0.8459	**0.8158**
		Improve		−12.60%	−4.09%	0.18%	0.17%	−5.72%	0.18%	3.74%
		RMSE	1.0796	1.3138	1.2106	1.0772	1.0773	1.1272	1.0772	**1.0369**
		Improve		−21.69%	−12.13%	0.22%	0.21%	−4.41%	0.22%	3.95%
	80%	MAE	0.8528	0.9621	0.8931	0.8527	0.8528	0.8907	0.8526	**0.8206**
		Improve		−12.82%	−4.72%	0.01%	0.00%	−4.44%	0.01%	3.78%
		RMSE	1.0850	1.3255	1.2304	1.0849	1.0850	1.1195	1.0848	**1.0413**
		Improve		−22.17%	−13.40%	0.01%	0.00%	−3.18%	0.02%	4.03%
	70%	MAE	0.8576	0.9706	0.9062	0.8575	0.8576	0.8976	0.8575	**0.8250**
		Improve		−13.17%	−5.67%	0.01%	0.00%	−4.66%	0.01%	3.80%
		RMSE	1.0894	1.3395	1.2547	1.0936	1.0894	1.1313	1.0894	**1.0461**
		Improve		−22.96%	−15.17%	−0.39%	0.00%	−3.85%	0.00%	3.97%

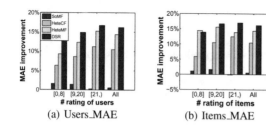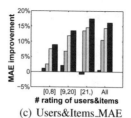

(a) Users_MAE (b) Items_MAE (c) Users&Items_MAE

Fig. 5.11 MAE improvement against PMF on various cold-start levels

of rated movies (e.g., [0, 8] denotes users rated no more than 8 movies and "All" means all users in Fig. 5.11). We conduct similar experiments on cold-start items and users&items (users and items are both cold-start). The experiments are shown in Fig. 5.11. Once again, we find that 3 HIN-based methods all are effective for cold-start users and items. Moreover, DSR always has the highest MAE improvement on almost all conditions, due to dual similarity regularization on users and items. It is reasonable since the DSR method takes much constraint information of users and items into account which would play a crucial role when there is little available information of users or items. The more detailed method description and experiment validation can be seen in [3, 26].

5.4 Conclusions

In recent years, recommendation has become a very popular application to alleviate information overload, and many recommendation techniques have been proposed. Recommender system includes a lot of object types and the rich relations among object types, so we can naturally constitute a heterogeneous information network from recommended system. The comprehensive information integration and rich semantic information of HIN make it promising to generate better recommendation. In this chapter, we introduce two types of recommendation methods with HIN. One type of methods employ the semantic path-based similarity measure to recommend items directly, and the other type of methods utilize rich similarity generated by meta paths to extend traditional matrix factorization methods. Experiments not only validate the effectiveness of these proposed methods but also show the benefits of information integration with heterogeneous network. In the future work, we can exploit the power and benefits of information integration with heterogeneous network in more applications.

References

1. BellogíN, R., Cantador, I., Castells, P.: A comparative study of heterogeneous item recommendations in social systems. Inf. Sci. **221**, 142–169 (2013)
2. Cantador, I., Bellogin, A., Vallet, D.: Content-based recommendation in social tagging systems. In: RecSys, pp. 237–240 (2010)
3. Cao, X., Zheng, Y., Shi, C., Li, J., Wu, B.: Link prediction in Schema-Rich heterogeneous information network. In: PAKDD, pp. 449–460 (2016)
4. Jamali, M., Ester, M.: Trustwalker: a random walk model for combining trust-based and item-based recommendation. In: KDD, pp. 397–406 (2009)
5. Jamali, M., Lakshmanan, L.V.S.: HeteroMF: recommendation in heterogeneous information networks using context dependent factor models. In: WWW, pp. 643–654 (2013)
6. Lao, N., Cohen, W.: Fast query execution for retrieval models based on path constrained random walks. In: KDD, pp. 881–888 (2010)
7. Lin, C.J.: Projected gradient methods for non-negative matrix factorization. In: Neural Computation, pp. 2279–2756 (2007)
8. Lippert, C., Weber, S.H., Huang, Y., Tresp, V., Schubert, M., Kriegel, H.P.: Relation prediction in multi-relational domains using matrix factorization. In: NIPS Workshop on Structured Input Structure Output (2008)
9. Luo, C., Pang, W., Wang, Z.: Hete-CF: social-based collaborative filtering recommendation using heterogeneous relations. In: ICDM, pp. 917–922 (2014)
10. Ma, H., Yang, H., Lyu, M.R., King, I.: SoRec: social recommendation using probabilistic matrix factorization. In: CIKM, pp. 931–940 (2008)
11. Ma, H., King, I., Lyu, M.R.: Learning to recommend with social trust ensemble. In: SIGIR, pp. 203–210 (2009)
12. Ma, H., Zhou, T.C., Lyu, M.R., King, I.: Improving recommender systems by incorporating social contextual information. ACM Trans. Inf. Syst. **29**(2), 9 (2011)
13. Ma, H., Zhou, D., Liu, C., Lyu, M.R., King, I.: Recommender systems with social regularization. In: WSDM, pp. 287–296 (2011)
14. Salakhutdinov, R., Mnih, A., Salakhutdinov, R., Mnih, A.: Probabilistic matrix factorization. In: NIPS **20**, 1257–1264 (2008)
15. Shardanand, U., Maes, P.: Social information filtering: algorithms for automating word of mouth. In: Sigchi Conference on Human Factors in Computing Systems, pp. 210–217 (1995)
16. Shi, C., Kong, X., Huang, Y., Philip, S.Y., Wu, B.: Hetesim: a general framework for relevance measure in heterogeneous networks. IEEE Trans. Knowl. Data Eng. **26**(10), 2479–2492 (2014)
17. Shi, C., Liu, J., Zhuang, F., Yu, P.S., Wu, B.: Integrating heterogeneous information via flexible regularization framework for recommendation. Knowl. Inf. Syst. **49**(3), 1–25 (2015)
18. Srebro, N., Jaakkola, T.: Weighted low-rank approximations. In: ICML, pp. 720–727 (2003)
19. Sun, Y.Z., Han, J.W., Yan, X.F., Yu, P.S., Wu, T.: PathSim: meta path-based top-K similarity search in heterogeneous information networks. In: VLDB, pp. 992–1003 (2011)
20. Sun, Y., Han, J.: Mining heterogeneous information networks: a structural analysis approach. SIGKDD Explor. **14**(2), 20–28 (2012)
21. Yang, X., Steck, H., Liu, Y.: Circle-based recommendation in online social networks. In: KDD, pp. 1267–1275 (2012)
22. Yu, X., Ren, X., Gu, Q., Sun, Y., Han, J.: Collaborative filtering with entity similarity regularization in heterogeneous information networks. In: IJCAI-HINA Workshop (2013)
23. Yu, X., Ren, X., Sun, Y., Sturt, B., Khandelwal, U., Gu, Q., Norick, B., Han, J.: Recommendation in heterogeneous information networks with implicit user feedback. In: RecSys, pp. 347–350 (2013)
24. Yu, X., Ren, X., Sun, Y., Gu, Q., Sturt, B., Khandelwal, U., Norick, B., Han, J.: Personalized entity recommendation: a heterogeneous information network approach. In: WSDM, pp. 283–292 (2014)
25. Yuan, Q., Chen, L., Zhao, S.: Factorization vs. regularization: fusing heterogeneous social relationships in top-n recommendation. In: RecSys, pp. 245–252 (2011)

26. Zheng, J., Liu, J., Shi, C., Zhuang, F., Li, J., Wu, B.: Dual similarity regularization for recommendation. In: PAKDD, pp. 542–554 (2016)

Chapter 6
Fusion Learning on Heterogeneous Social Networks

Jiawei Zhang

Abstract Looking from a global perspective, the landscape of online social networks is highly fragmented. A large number of online social networks have appeared, which can provide the users with various types of services. Generally, the information available in the these online social networks is of diverse categories, which can be represented as heterogeneous information networks (HIN) formally. Meanwhile, in such an age of online social media, users usually participate in multiple online social networks simultaneously to enjoy more social networks services, who can act as bridges connecting different networks together. So multiple HINs not only represent information in single network, but also fuse information from multiple networks. Formally, the online social networks sharing common users are named as the aligned social networks, and these shared users who act like anchors aligning the networks are called the anchor users. The heterogeneous information generated by users' social activities in the multiple aligned social networks provides social network practitioners and researchers with the opportunities to study individual user's social behaviors across multiple social platforms simultaneously.

6.1 Network Alignment

6.1.1 Overview

Heterogeneous information networks (HIN) is a very general network representation in the real world and lots of network structured data can be represented as HINs formally, such as collaboration networks, online social networks, and knowledge base. Meta path first proposed by Sun et al. for heterogeneous information networks in [32] is a powerful tool, which can be applied in link prediction problems [31, 34], clustering problems [32, 33], searching and ranking problems [16, 37], and collective classification problem [11] in HINs. However, most of these applications are within one single network only, meta path extracted from which are called the intra-network meta path.

Meanwhile, to enjoy the social network services from multiple online social networks simultaneously, users nowadays are usually involved in multiple online social

© Springer International Publishing AG 2017

C. Shi and P.S. Yu, *Heterogeneous Information Network Analysis and Applications*, Data Analytics, DOI 10.1007/978-3-319-56212-4_6

networks at the same time. Formally, the online social networks sharing common users are named as the aligned social networks, and these shared users who act like anchors aligning the networks are called the anchor users. Social activity analysis across aligned social networks has become a hot research topic in recent years and many pioneer works have been done on this topic. Zhang et al. propose to study the network alignment problem between pairwise fully aligned networks [12], pairwise partially aligned networks [44, 45, 47], and multiple partially aligned networks [46].

Based on the aligned networks, various kinds of application problems have been studied across multiple social platforms, including friend recommendation and social link prediction for new users [42] and emerging networks [43, 44, 50], location recommendation [43], community detection for emerging networks [48] and synergistic clustering across networks [9, 28, 36], information diffusion [39, 40], viral marketing [39], and tipping user identification [40]. To handle the heterogeneous information available across the aligned social networks, the meta path concept is firstly extended to inter-network scenario [45, 50] and applied to address various synergistic knowledge discovery problems across partially aligned heterogeneous social networks, which include network alignment [45], link recommendation [50], community detection [36], and information diffusion [39, 40].

Network alignment problem has been well studied in bioinformatics, e.g., protein-protein interaction (PPI) network alignment [10, 14, 17, 30]. Most network alignment approaches focus on finding approximate isomorphism between two graphs [10, 14, 30]. Because of the intractability of the problem, existing methods usually rely on practical heuristics to solve the problem [10, 17]. Meanwhile, in recent years, some works have been done on aligning social networks [12, 13, 22]. Various network alignment models have been proposed to address the problem, which include the supervised classification-based network alignment methods [12, 45], PU (positive and unlabeled) classification-based method [44], and unsupervised matrix estimation-based methods [46, 47].

In this chapter, we will take heterogeneous social network as an example and introduce the network alignment problem and UNICOAT model studied in [47]. In the network alignment problem, we aim at identifying the common users' accounts (i.e., the anchor links) across different social platforms based on the heterogeneous information available in the networks, which includes both the network structure information and various types of attribute information.

6.1.2 Terminology Definition and Social Meta Path

Before introducing the proposed framework for the network alignment problem, we will first introduce a set of terminologies that will be used both in this section and throughout this chapter, including heterogeneous information networks, multiple aligned social networks, anchor links, and the intra-network meta path and inter-network meta path. A set of intra-network and inter-network meta paths will also be

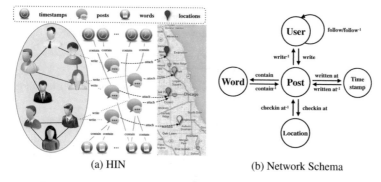

Fig. 6.1 An example of HIN and the corresponding network schema

introduced, whose notations, representation, and physical meanings will be illustrated as follows.

6.1.2.1 Terminology Definition

As shown in Fig. 6.1a, online social networks usually contain heterogeneous information involving different types of nodes, e.g., users, posts, words, time stamps, and location checkins, as well as complex links among the nodes, e.g., friendship links among users, write links between users and posts, and the contain/attach links between posts and words, time stamps, and checkins. Formally, such a kind of online social network can be represented as the heterogeneous information networks.

Definition 6.1 (*Heterogeneous Information Networks*) A **heterogeneous information network** can be represented as $G = (V, E)$, where the nodes in set $V = \bigcup_i V_i$ and the links in set $E = \bigcup_i E_i$ are of different categories, respectively.

Users nowadays are usually involved in multiple online social networks simultaneously to enjoy more social network services. Formally, the online social networks sharing common users can be defined as the multiple aligned social networks [12], which are connected by the anchor links [42] between the accounts of shared users, i.e., the anchor users [50].

Definition 6.2 (*Multiple Aligned Social Networks*) The **multiple aligned social networks** can be represented as $G = (\{G^i\}_i, \{A^{(i,j)}\}_{i,j})$, where $G^i = (V^i, E^i)$ denotes the i_{th} **heterogeneous information network** and $A^{(i,j)}$ represents the set of undirected anchor links between networks G^i and G^j.

Definition 6.3 (*Anchor Link*) Between networks G^i and G^j, the set of undirected anchor links $A^{(i,j)}$ can be represented as $A^{(i,j)} = \{(u^i_m, v^j_n)|u^i_m \in U^i, v^j_n \in U^j, u^i_m \text{ and } v^j_n \text{ are the accounts of the same user}\}$, where $U^i \subset V^i$ and $U^j \subset V^j$ are the user node sets in networks G^i and G^j, respectively.

Table 6.1 Summary of intra-network social meta paths

ID	Notation	Intra-network social meta path	Semantics
1	$U \rightarrow U$	User $\xrightarrow{follow}$ User	Follow
2	$U \rightarrow U \rightarrow U$	User $\xrightarrow{follow}$ User $\xrightarrow{follow}$ User	Follower of follower
3	$U \rightarrow U \leftarrow U$	User $\xrightarrow{follow}$ User $\xleftarrow{follow}$ User	Common out-neighbor
4	$U \leftarrow U \rightarrow U$	User $\xleftarrow{follow}$ User $\xrightarrow{follow}$ User	Common in-neighbor
5	$U \rightarrow P \rightarrow W \leftarrow P \leftarrow U$	User $\xrightarrow{write}$ Post $\xrightarrow{contain}$ Word $\xleftarrow{contain}$ Post $\xleftarrow{write}$ User	Posts containing common words
6	$U \rightarrow P \rightarrow T \leftarrow P \leftarrow U$	User $\xrightarrow{write}$ Post $\xrightarrow{contain}$ Time $\xleftarrow{contain}$ Post $\xleftarrow{write}$ User	Posts containing common time stamps
7	$U \rightarrow P \rightarrow L \leftarrow P \leftarrow U$	User $\xrightarrow{write}$ Post $\xrightarrow{attach}$ Location $\xleftarrow{attach}$ Post $\xleftarrow{write}$ User	Posts attaching common location check-ins

One way to model the heterogeneous information available across the multiple aligned social networks is meta path [33, 36, 50], which abstracts the connections among the different categories of nodes as sequences of link types connected by the node types. For instance, given the social network with its schema shown in Fig. 6.1, a summary of the intra-network social meta paths extracted from the network is provided in Table 6.1.

Definition 6.4 (*Intra-Network Meta Path*) Given a **heterogeneous information network** $G^i = (V^i, E^i)$, we can represents its **networks schema** as $S(G^i) = (T^i, R^i)$, where T^i denotes the types of nodes in V^i and R^i denotes the types of links in E^i. Formally, based on the **network schema**, we can define the **meta path** as a sequence $P : T_1^i \xrightarrow{R_1^i} T_2^i \xrightarrow{R_2^i} \cdots \xrightarrow{R_m^i} T_{m+1}^i$, where $T_m^i \in T^i$ and $R_n^i \in R^i$ are the node and link types available in network G^i, respectively.

Besides the intra-network meta Paths, via the anchor links and other shared information entities, nodes across different networks can also get connected by the inter-network meta paths.

Definition 6.5 (*Inter-Network Meta Path*) Given a meta path P consisting of sequences of link types, P is an **inter-network meta path** between networks G^i and G^j iff P involves the node types and link types from the schema of both network G^i and network G^j.

Table 6.2 Summary of inter-network social meta paths

ID	Notation	Intra-network social meta path	Semantics
1	$U^i \rightarrow U^i \leftrightarrow U^j \leftarrow U^j$	$\text{User}^i \xrightarrow{follow} \text{User}^i \xleftrightarrow{Anchor}$ $\text{User}^j \xleftarrow{follow} \text{User}^j$	Inter-network common out-neighbor
2	$U^i \leftarrow U^i \leftrightarrow U^j \rightarrow U^j$	$\text{User}^i \xleftarrow{follow} \text{User}^i \xleftrightarrow{Anchor}$ $\text{User}^j \xrightarrow{follow} \text{User}^j$	Inter-network common in-neighbor
3	$U^i \rightarrow U^i \leftrightarrow U^j \rightarrow U^j$	$\text{User}^i \xrightarrow{follow} \text{User}^i \xleftrightarrow{Anchor}$ $\text{User}^j \xrightarrow{follow} \text{User}^j$	Inter-network common out in-neighbor
4	$U^i \leftarrow U^i \leftrightarrow U^j \leftarrow U^j$	$\text{User}^i \xleftarrow{follow} \text{User}^i \xleftrightarrow{Anchor}$ $\text{User}^j \xleftarrow{follow} \text{User}^j$	Inter-network common in out-neighbor
5	$U^i \rightarrow P^i \rightarrow L \leftarrow P^j \leftarrow U^j$	$\text{User}^i \xrightarrow{write} \text{Post}^i \xrightarrow{checkin\ at}$ $\text{Location} \xleftarrow{checkin\ at} \text{Post}^j$ $\xleftarrow{write} \text{User}^j$	Inter-network common location checkins
7	$U^i \rightarrow P^i \rightarrow T \leftarrow P^j \leftarrow U^j$	$\text{User}^i \xrightarrow{write} \text{Post}^i \xrightarrow{at} \text{Time}$ $\xleftarrow{at} \text{Post}^j \xleftarrow{write} \text{User}^j$	Inter-network common time stamps
8	$U^i \rightarrow P^i \rightarrow W \leftarrow P^j \leftarrow U^j$	$\text{User}^i \xrightarrow{write} \text{Post}^i \xrightarrow{contain}$ $\text{Word} \xleftarrow{contain} \text{Post}^j \xleftarrow{write}$ User^j	Inter-network common words

The simplest inter-network meta path between networks G^i and G^j will be the anchor meta path [45, 50] involving the user node types from G^i and G^j and the anchor link type between G^i and G^j. Some inter-network meta path examples are summarized in Table 6.2.

6.1.2.2 Social Meta Paths

Meta paths can actually connect various categories of node types from the network, and those starting and ending with user node types are formally named as the social meta paths [36] specifically. In this chapter, we will use the Foursquare and Twitter networks as the example of multiple aligned social networks, which actually share a large amount of common users. As shown in Fig. 6.1a, both the Foursquare and Twitter networks can be represented as a heterogeneous information network $G = (V, E)$, where the node set $V = U \cup P \cup L \cup T \cup W$ involves the nodes of users, posts, locations, time stamps, and words, while the link set $E = E_{u,u} \cup E_{u,p} \cup E_{p,l} \cup E_{p,t} \cup E_{p,w}$ contains the links among users, between users and posts, and those between posts and locations, time stamps, and words, respectively. The corresponding network schema of the HIN is shown in Fig. 6.1b. Based on the network schema, a set of intra-network social meta paths can be extracted and defined from the network, which are shown in Table 6.1.

Besides the intra-network social metapaths, in Table 6.2, we also show a list of inter-network social meta paths connecting user node types in networks G^i and G^j, respectively. These inter-network social meta paths connect user nodes across networks via either the anchor links or other common information entities, e.g., location checkins, words, and time stamps.

6.1.3 Cross-Network Network Alignment

Formally, given networks $G^1, G^2, \cdots, G^n$ together with information available in them, the network alignment problem aims at identifying the anchor link sets $A^{(1,2)}, A^{(1,3)}, \cdots, A^{(n-1,n)}$ between pairwise networks. The set of anchor links to be inferred between networks G^i and G^j can be represented as $A^{(i,j)}$, which aligns users between networks G^i and G^j. Considering that users in different social networks are associated with both links and attribute information, the quality of the inferred anchor links $A^{(i,j)}$ can be measured by the costs introduced by such mappings calculated with users' link and attribute information, i.e.,

$$cost(A^{(i,j)}) = \text{cost in links } (A^{(i,j)}) + \alpha \cdot \text{cost in attributes}(A^{(i,j)}), \qquad (6.1)$$

where α denotes the weight of the cost obtained from the attribute information.

6.1.3.1 Structure Information-Based Network Alignment

Based on the social links among users in both G^i and G^j (i.e., $E^i_{u,u}$ and $E^j_{u,u}$, respectively), we can construct the binary social adjacency matrices $\mathbf{S}^i \in \mathbb{R}^{|U^i| \times |U^i|}$ and $\mathbf{S}^j \in \mathbb{R}^{|U^j| \times |U^j|}$ for networks G^i and G^j, respectively. Entries in $\mathbf{S}^i$ and $\mathbf{S}^j$ (e.g., $\mathbf{S}^i(p,q)$ and $\mathbf{S}^j(l,m)$) will be assigned with value 1 iff the corresponding social links (u^i_p, u^i_q) and (u^j_l, u^j_m) exist in G^i and G^j, where $u^i_p, u^i_q \in U^i$ and $u^j_l, v^j_m \in U^j$ are users in networks G^i and G^j.

Via the inferred anchor links $A^{(i,j)}$, users as well as their social connections can be mapped between networks G^i and G^j. We can represent the inferred anchor links $A^{(i,j)}$ with binary user transitional matrix $\mathbf{P} \in \mathbb{R}^{|U^i| \times |U^j|}$, where the (i_{th}, j_{th}) entry $\mathbf{P}(p,q) = 1$ iff link $(u^i_p, u^j_q) \in A^{(i,j)}$. Considering that the constraint on anchor links is one-to-one, each column and each row of $\mathbf{P}$ can contain at most one entry being assigned with value 1, i.e.,

$$\mathbf{P}\mathbf{1}^{|U^j| \times 1} \leq \mathbf{1}^{|U^i| \times 1}, \ \mathbf{P}^\top \mathbf{1}^{|U^i| \times 1} \leq \mathbf{1}^{|U^j| \times 1}, \qquad (6.2)$$

where $\mathbf{P}\mathbf{1}^{|U^j| \times 1}$ and $\mathbf{P}^\top \mathbf{1}^{|U^i| \times 1}$ can get the sum of rows and columns of matrix $\mathbf{P}$, respectively. Eq. $\mathbf{P}\mathbf{1}^{|U^j| \times 1} \leq \mathbf{1}^{|U^i| \times 1}$ denotes that every entry of the left vector is no greater than the corresponding entry in the right vector.

Matrix $\mathbf{P}$ is an equivalent representation of anchor link set $A^{(i,j)}$. Next, we will infer the optimal user transitional matrix $\mathbf{P}$, from which we can obtain the optimal anchor link set $A^{(i,j)}$.

The optimal anchor links are those which can minimize the inconsistency of mapped social links across networks and the cost introduced by the inferred anchor link set $A^{(i,j)}$ with the link information can be represented as

$$\text{cost in link}(A^{(i,j)}) = \text{cost in link}(\mathbf{P}) = \left\| \mathbf{P}^\top \mathbf{S}^i \mathbf{P} - \mathbf{S}^j \right\|_F^2, \tag{6.3}$$

where $\|\cdot\|_F$ denotes the Frobenius norm of the corresponding matrix and $\mathbf{P}^\top$ is the transpose of matrix $\mathbf{P}$.

6.1.3.2 Attribute Information-Based Network Alignment

With these different attribute information (i.e., username, temporal activity, and text content), we can calculate the similarities between users across networks G^i and G^j based on the inter-network social meta paths. To measure the social closeness among users across directed heterogeneous information networks, we propose a new closeness measure named INMP-Sim (Inter-Network Meta Path-based Similarity) as follows.

Definition 6.6 (*INMP-Sim*) Let $P_i(x \rightsquigarrow y)$ and $P_i(x \rightsquigarrow \cdot)$ be the sets of path instances of **inter-network meta paths** $\# i$ going from x to y and those going from x to other nodes in the network. The INMP-Sim of node pair (x, y) is defined as

$$\text{INMP-Sim}(x, y) = \sum_i \omega_i \left(\frac{|P_i(x \rightsquigarrow y)| + |P_i(y \rightsquigarrow x)|}{|P_i(x \rightsquigarrow \cdot)| + |P_i(y \rightsquigarrow \cdot)|} \right), \tag{6.4}$$

where ω_i is the weight of **inter-network meta paths** $\# i$ and $\sum_i \omega_i = 1$.

Formally, we represent such similarity matrix as $\Lambda \in \mathbb{R}^{|U^i| \times |U^j|}$, where entry $\Lambda(p, q)$ is the similarity between u_p^i and u_q^j. Similar users across social networks are more likely to be the same user and anchor links $A_u^{(i,j)}$ that align similar users together should lead to lower cost. In this chapter, the cost function introduced by the inferred anchor links $A_u^{(i,j)}$ in attribute information is represented as

$$\text{cost in attribute}(A_u^{(i,j)}) = \text{cost in attribute}(\mathbf{P}) = - \|\mathbf{P} \circ \Lambda\|_1, \tag{6.5}$$

where $\|\cdot\|_1$ is the L_1 norm of the corresponding matrix, entry $(\mathbf{P} \circ \Lambda)(i, l)$ can be represented as $P(i, l) \cdot \Lambda(i, l)$, and $\mathbf{P} \circ \Lambda$ denotes the Hadamard product of matrices $\mathbf{P}$ and Λ.

6.1.3.3 Joint Objective Function for Network Alignment

Both link and attribute information is important for anchor link inference. By taking these two categories of information into consideration simultaneously, we can represent the optimal user transitional matrix $\mathbf{P}^*$ which can lead to the minimum cost as follows:

$$
\begin{aligned}
\mathbf{P}^* &= \arg\min_{\mathbf{P}} cost(\mathbb{A}_u^{(i,j)}) \\
&= \arg\min_{\mathbf{P}} \left\| \mathbf{P}^\top \mathbf{S}^i \mathbf{P} - \mathbf{S}^j \right\|_F^2 - \alpha \cdot \| \mathbf{P} \circ \Lambda \|_1 \qquad (6.6) \\
s.t. \mathbf{P} &\in \{0, 1\}^{|\mho^i| \times |\mho^j|}, \\
&\mathbf{P}\mathbf{1}^{|\mho^j| \times 1} \le \mathbf{1}^{|\mho^i| \times 1}, \mathbf{P}^\top \mathbf{1}^{|\mho^i| \times 1} \le \mathbf{1}^{|\mho^j| \times 1}.
\end{aligned}
$$

The objective function is an constrained $0 - 1$ integer programming problem, which is hard to address mathematically. Many relaxation algorithms have been proposed so far. For more information about how to resolve the objective function, please refer to [47].

6.1.4 Experiments

To test the effectiveness of the proposed UNICOAT model, in this section, extensive experiments have been done on two real-world partially co-aligned online social networks: Foursquare and Twitter.

6.1.4.1 Dataset

The social networks dataset used in this chapter are Foursquare and Twitter, which are co-aligned by both users and locations shared between these two networks. These two social network datasets are crawled during November, 2012, whose statistical information is available in Table 6.3. More detailed descriptions and the crawling method is available in [43, 50].

To show the advantages of UNICOAT in addressing the NETWORK ALIGNMENT problem, we compare UNICOAT with many different baseline methods. Considering that no known anchor links are available actually in the NETWORK ALIGNMENT problem, as a result, no existing supervised network alignment methods (e.g., MNA [12]) can be applied. All the comparison methods are based on unsupervised learning settings, which can be divided into 4 categories:

Table 6.3 Properties of the heterogeneous networks

	Property	Network	
		Twitter	Foursquare
# node	User	5,223	5,392
	Tweet/tip	9,490,707	48,756
	Location	297,182	38,921
# link	Friend/follow	164,920	76,972
	Write	9,490,707	48,756
	Locate	615,515	48,756

Co-Alignment Methods

- UNICOAT: Method UNICOAT can align two online social networks based on the shared users and locations simultaneously, which consists of two steps: (1) unsupervised potential anchor links inference; (2) co-matching of social networks to prune redundant anchor links to maintain the one-to-one constraint.

Bipartite Graph Alignment Methods

- BIGALIGN: Method BIGALIGN is a bipartite network alignment method introduced in [13], which can align two bipartite graphs (e.g., user-product bipartite graph) simultaneously with link information only.
- BIGALIGNEXT: Method BIGALIGNEXT is a bipartite network alignment method. BIGALIGNEXT can align user-location bipartite networks with both location links between users and locations as well as attribute information about users and locations across networks.

Isolated Alignment Methods

- ISO: Method ISO is an unsupervised network alignment method introduced in [13]. ISO merely infers the anchor links only based on the friendship information among users.
- ISOEXT: Method ISOEXT is an unsupervised network alignment method, which is identical to ISO but utilizes both friendship links among users and attribute information of users.

Traditional Unsupervised Link Prediction Methods

- Relative Degree Distance-based Network Alignment: RDD is the heuristics-based unsupervised network alignment method introduced in [13] to fill in the initial values of the cross-network transitional matrices, e.g., $\mathbf{P}$. For any two users/location $u_l^{(i)}$ and $u_m^{(j)}$ in networks $G^{(i)}$ and $G^{(j)}$, the relative degree distance between them can be represented as $RDD(u_l^{(i)}, u_m^{(j)}) = \left(1 + \frac{|deg(u_l^{(i)}) - deg(u_m^{(j)})|}{(deg(u_l^{(i)}) + deg(u_m^{(j)}))/2}\right)^{-1}$. High relative degree distance denotes lower confidence score of anchor link $(u_l^{(i)}, u_m^{(j)})$.

Methods UNICOAT (the first step), BIGALIGN, BIGALIGNEXT ISO, ISOEXT and RDD can output the confidence scores of potential inferred links but no labels are available, whose performance can be evaluated by metrics such as AUC and Precision@100. As to method UNICOAT, links selected finally in the matching are assumed to achieve confidence score 1.0 and label $+1$, while the remaining can achieve confidence score 0.0 and label -1. As a result, UNICOAT can also output the labels of potential anchor links, whose performance can be evaluated by various metrics, e.g., AUC, Precision@100, Precision, Recall, F1, and Accuracy, simultaneously.

The experiment results of addressing the NETWORK ALIGNMENT problem are available in Table 6.4 and Fig. 6.2. In Fig. 6.2, we fix $\theta = 1$ and show the results achieved by comparison methods without matching step (i.e., methods UNICOAT (the first step), BIGALIGN, BIGALIGNEXT, ISO, ISOEXT and RDD) evaluated by AUC and Precision@100. Methods ISO and ISOEXT can only be applied to align networks via

Table 6.4 Performance comparison of different methods for inferring user anchor links (UNICOAT here denotes the first step of UNICOAT only)

Measure		θ				
	Methods	1	2	3	4	5
AUC	UNICOAT	**0.868**	**0.831**	**0.814**	**0.804**	**0.799**
	BIGALIGNEXT	0.813	0.779	0.759	0.752	0.749
	BIGALIGN	0.568	0.557	0.555	0.552	0.550
	ISOEXT	0.818	0.782	0.762	0.754	0.61
	ISO	0.547	0.529	0.52	0.518	0.516
	RDD	0.531	0.530	0.523	0.514	0.508
Prec@100	UNICOAT	**0.705**	**0.688**	**0.657**	**0.640**	**0.556**
	BIGALIGNEXT	0.587	0.507	0.472	0.434	0.327
	BIGALIGN	0.347	0.284	0.265	0.228	0.220
	ISOEXT	0.427	0.391	0.373	0.352	0.301
	ISO	0.301	0.253	0.225	0.216	0.208
	RDD	0.234	0.228	0.207	0.172	0.127

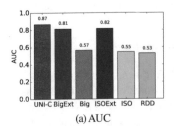
(a) AUC

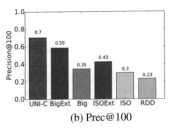
(b) Prec@100

Fig. 6.2 Performance of methods without matching in inferring anchor links (UNICOAT here denotes the first step of UNICOAT only)

user generated information. In Fig. 6.2, we can observe that (1) UNICOAT performs the best among all the comparison methods in inferring anchor links evaluated by both AUC and Precision@100. For example, in Fig. 6.2, UNICOAT can achieve AUC score of 0.87, which is over 6% better than BIGALIGNEXT and ISOEXT, and 50% higher than the AUC score achieved by BIGALIGN, ISO and RDD. Similar performance of UNICOAT is available in other plots. It demonstrates that utilizing the heterogeneous information in the network to infer anchor links simultaneously can improve the results a lot. (2) BIGALIGNEXT and ISOEXT can achieve better performance than BIGALIGN and ISO. Recalling that methods BIGALIGNEXT and ISOEXT use both the link and attribute information, while BIGALIGN and ISO use the link information. It justifies that the attribute information of both users is helpful for inferring anchor links across networks. (3) By comparing UNICOAT with RDD (i.e., the initialization method of matrices $\mathbf{P}$ in UNICOAT), we observe that UNICOAT can outperform RDD with significant advantages. It proves the effectiveness of the proposed network co-alignment model, which can obtain better results than the initial value.

6.1.4.2 Sensitivity Analysis

In Fig. 6.2, parameter θ is fixed as 1. In Table 6.4, we further change it with values in $\{1, 2, 3, 4, 5\}$ by adding more non-anchor users into the network. Generally, with more non-anchor users, the NETWORK ALIGNMENT will become more difficult and the performance of all the methods will degrade, but UNICOAT can achieve the best performance consistently. For example, when $\theta = 5$, the AUC score achieved by UNICOAT in inferring social links is 0.799, which is 6.7, 45, 31, 54.8, and 57.2% higher than that gained by BIGALIGNEXT, BIGALIGN, ISOEXT, ISO, and RDD, respectively. Similar observations can be obtained from the anchor links inference results evaluated by Precision@100 in Table 6.4.

In the previous part, we have shown the performance of methods without matching step, while anchor links inferred by which cannot meet the one-to-one constraint. Next, we will test the effectiveness of the matching step in pruning the non-existing anchor links and the results achieved by UNICOAT (the second step) are shown in Fig. 6.3. Parameter θ are assigned with values in $\{1, 2, 3, 4, 5\}$. The anchor links inferred by UNICOAT can all meet the one-to-one constraint and are of high quality. For example, when $\theta = 1$, the Precision, Recall, F1, and Accuracy achieved by UNICOAT are 0.73, 0.54, 0.62, and 0.75, respectively, in inferring anchor links. As θ increases, Recall and F1 scores achieved by UNICOAT will decrease as it will be more hard to identify the real anchor links among larger number of potential ones. Meanwhile, the Precision and Accuracy of UNICOAT will increase. The potential reason can be due to the class imbalance problem. By adding more non-anchor users to the network, more non-existing anchor links (i.e., the negative class links) will be introduced and UNICOAT can achieve higher Precision and Accuracy by predicting more negative instances correctly.

Fig. 6.3 Performance of
methods with matching in
inferring anchor links
(UNICOAT here includes
both two steps of UNICOAT)

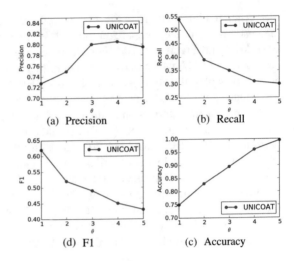

(a) Precision

(b) Recall

(d) F1

(c) Accuracy

6.2 Link Transfer Across Aligned Networks

To investigate users' social activities and the propagation of information across different social platforms, several application problems will also be introduce in this chapter after aligning the networks. One important work will be the link prediction problems, which aims at infer potential connections among the information entities in the networks. Link prediction across the multiple aligned social networks is not an easy task, and the heterogeneity of the social networks renders the problem more challenging to solve.

6.2.1 Overview

Link prediction in social networks first proposed by Liben-Nowell [18] has been a hot research topic and many different methods have been proposed. Liben-Nowell [18] proposes many unsupervised link predicators to predict the social connections among users. Later, Hasan [1] proposes to predict links by using supervised learning methods. An extensive survey of link prediction works is available in [7, 8]. Most existing link prediction works are based on one single network but many researchers start to shift their attention to multiple networks. Dong et al. [5] propose to do link prediction with multiple information sources. Zhang et al. introduce the link prediction problem across aligned networks for new users [42] and emerging networks [43, 44] based on supervised classification models [42] and PU classification models [43, 44], respectively. Depending on the specific application settings, the links to be predicted are usually subject to different cardinality constraints, like one-to-one [12], one-to-many [49], and many-to-many [50]. For links with each type of the cardinality constraints, different link prediction models have been proposed already. Zhang et al. propose to unify these different link prediction tasks into a general link prediction problem and introduce a general model for the problem [41].

In this chapter, we will briefly introduce the multinetwork synergistic PU link prediction framework MLI as follows. Given a network screenshot, MLI labels the existing and non-existing social links among users as positive and unlabeled instances, respectively, where the unlabeled links involve both positive and negative links at the same time. Therefore, the link prediction task can be transferred into a PU learning task.

6.2.2 Cross-Network Link Prediction

Formally, given multiple aligned networks $G = (\{G^1, G^2, \cdots, G^n\}, \{A^{(1,2)}, A^{(1,3)}, \cdots, A^{(n-1,n)}\})$, the objective of the cross-network link prediction problem is to infer the potential social connections which will be formed in the near future in networks $G^1, G^2, \cdots, G^n$, respectively.

6.2.2.1 PU Link Prediction Feature Extraction

Meta paths introduced in the previous sections can actually cover a large number of path instances connecting users across the network. Formally, we denote that node n (or link l) is an instance of node type T (or link type R) in the network as $n \in T$ (or $l \in R$). Identity function $I(a, A) = \begin{cases} 1, & \text{if } a \in A \\ 0, & \text{otherwise,} \end{cases}$ can check whether node/link a is an instance of node/link type A in the network. To consider the effect of the unconnected links when extracting features for social links in the network, we formally define the **Social Meta Path-based Features** to be:

Definition 6.7 (*Social Meta Path-based Features*) For a given link (u, v), the feature extracted for it based on meta path $P = T_1 \xrightarrow{R_1} T_2 \xrightarrow{R_2} \cdots \xrightarrow{R_{k-1}} T_k$ from the networks is defined to be the expected number of formed path instances between u and v across the networks:

$$x(u, v) = I(u, T_1)I(v, T_k) \sum_{n_1 \in \{u\}, n_2 \in T_2, \cdots, n_k \in \{v\}} \prod_{i=1}^{k-1} p(n_i, n_{i+1})I((n_i, n_{i+1}), R_i),$$

(6.7)

where $p(n_i, n_{i+1}) = 1.0$ if $(n_i, n_{i+1}) \in E_{u,u}$ and otherwise, $p(n_i, n_{i+1})$ denotes the **formation probability** of link (n_i, n_{i+1}) to be introduced in Sect. 6.2.2.3.

Based on the above social meta path-based feature definition and the extracted intra-network and inter-network meta paths, a set of features can be extracted for user pairs with the information across the aligned networks.

6.2.2.2 Meta Path-Based Feature Selection

Meanwhile, information transferred from aligned networks via the features extracted based on the inter-network social meta path can be helpful for improving link prediction performance in a given network but can be misleading as well, which is called the network difference problem. To solve the network difference problem, we propose to rank and select top K features from the feature vector extracted based on the intra-network and inter-network social meta paths, $\mathbf{x}$, from the multiple partially aligned heterogeneous networks.

Let variable $X_i \in \mathbf{x}$ be a feature extracted based on meta paths #i and variable Y be the label. $P(Y = y)$ denotes the prior probability that links in the training set having label y and $P(X_i = x)$ represents the frequency that feature X_i has value x. Information theory related measure mutual information (mi) is used as the ranking criteria:

$$mi(X_i) = \sum_x \sum_y P(X_i = x, Y = y) \log \frac{P(X_i = x, Y = y)}{P(X_i = x)P(Y = y)} \qquad (6.8)$$

Let $\bar{\mathbf{x}}$ be the features of the top K mi score selected from $\mathbf{x}$. In the next subsection, we will use the selected feature vector $\bar{\mathbf{x}}$ to build a novel PU link prediction model.

6.2.2.3 PU Link Prediction Method

As introduced at the beginning of this section, from a given network, e.g., G, we can get two disjoint sets of links: connected (i.e., formed) links P and unconnected links U. To differentiate these links, we define a new concept "connection state", z, to show whether a link is connected (i.e., formed) or unconnected in network G. For a given link l, if l is connected in the network, then $z(l) = +1$; otherwise, $z(l) = -1$. As a result, we can have the "connection states" of links in P and U to be: $z(P) = +1$ and $z(U) = -1$.

Besides the "connection state," links in the network can also have their own "labels," y, which can represent whether a link is to be formed or will never be formed in the network. For a given link l, if l has been formed or to be formed, then $y(l) = +1$; otherwise, $y(l) = -1$. Similarly, we can have the " labels" of links in P and U to be: $y(P) = +1$ but $y(U)$ can be either $+1$ or -1, as U can contain both links to be formed and links that will never be formed.

By using P and U as the positive and negative training sets, we can build a link connection prediction model M_c, which can be applied to predict whether a link exists in the original network, i.e., the connection state of a link. Let l be a link to be predicted, by applying M_c to classify l, we can get the connection probability of l to be:

Definition 6.8 (*Connection Probability*) The probability that link l's **connection states** is predicted to be **connected** (i.e., $z(l) = +1$) is formally defined as the **connection probability** of link l: $p(z(l) = +1|\bar{\mathbf{x}}(l))$.

Meanwhile, if we can obtain a set of links that "will never be formed", i.e., "-1" links, from the network, which together with P ("$+1$" links) can be used to build a link formation prediction model, M_f, which can be used to get the formation probability of l to be:

Definition 6.9 (*Formation Probability*) The probability that link l's **label** is predicted to be **formed or will be formed** (i.e., $y(l) = +1$) is formally defined as the **formation probability** of link l: $p(y(l) = +1|\bar{\mathbf{x}}(l))$.

However, from the network, we have no information about "links that will never be formed" (i.e., "-1" links). As a result, the formation probabilities of potential links that we aim to obtain can be very challenging to calculate. Meanwhile, the correlation between link l's connection probability and formation probability has been proved in existing works [6] to be:

$$p(y(l) = +1|\bar{\mathbf{x}}(l)) \propto p(z(l) = +1|\bar{\mathbf{x}}(l)). \tag{6.9}$$

In other words, for links whose connection probabilities are low, their formation probabilities will be relatively low as well. This rule can be utilized to extract links which can be more likely to be the reliable "-1" links from the network. We propose to apply the link connection prediction model M_c built with P and U to classify links in U to extract the reliable negative link set. Formally, such a kind of negative link extraction method is called the spy technique-based reliable negative link extraction. For more detailed information about method, please refer to [50].

With the extracted reliable negative link set RN, we can solve the PU link prediction problem with classification-based link prediction methods, where P and RN are used as the positive and negative training sets, respectively. Meanwhile, when applying the built model to predict links in L^i, the optimal labels, $\hat{\mathbf{Y}}^i$, of L^i, should be those which can maximize the following formation probabilities:

$$\hat{\mathbf{Y}}^i = \arg\max_{\mathbf{Y}^i} p(y(\mathrm{L}^i) = \mathbf{Y}^i | G^1, G^2, \cdots, G^k)$$

$$= \arg\max_{\mathbf{Y}^i} p(y(\mathrm{L}^i) = \mathbf{Y}^i | \bar{\mathbf{x}}(\mathrm{L}^i)) \tag{6.10}$$

where $y(\mathrm{L}^i) = \mathbf{Y}^i$ represents that links in L^i have labels $\mathbf{Y}^i$.

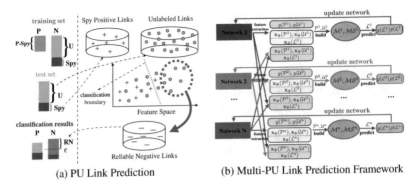

(a) PU Link Prediction   (b) Multi-PU Link Prediction Framework

Fig. 6.4 PU link prediction framework across multiple aligned networks

6.2.2.4 Multinetwork Link Prediction Framework

Method proposed in [50] is a general link prediction framework and can be applied to predict social links in n partially aligned networks simultaneously. When it comes to n partially aligned network, the optimal labels of potential links $\{L^1, L^2, \cdots, L^n\}$ of networks $G^1, G^2, \cdots, G^n$ will be:

$$\hat{Y}^1, \hat{Y}^2, \cdots, \hat{Y}^n$$
$$= \arg\max_{Y^1, Y^2, \dots, Y^n} p(y(L^1) = Y^1, y(L^2) = Y^2, \cdots, y(L^n) = Y^n | G^1, G^2, \cdots, G^n) \tag{6.11}$$

The above target function is very complex to solve and we propose to obtain the solution by updating one variable, e.g., Y^1, and fix other variables, e.g., $Y^2, \cdots, Y^n$, alternatively with the following equation [43]:

$$\begin{cases} (\hat{Y}^1)^{(\tau)} & = \arg\max_{Y^1} p(y(L^1) = Y^1 | G^1, \cdots, G^n, (\hat{Y}^2)^{(\tau-1)}, (\hat{Y}^3)^{(\tau-1)}, \cdots, (\hat{Y}^n)^{(\tau-1)}) \\ (\hat{Y}^2)^{(\tau)} & = \arg\max_{Y^2} p(y(L^2) = Y^2 | G^1, \cdots, G^n, (\hat{Y}^1)^{(\tau)}, (\hat{Y}^3)^{(\tau-1)}, \cdots, (\hat{Y}^n)^{(\tau-1)}) \\ \quad \dots\dots \\ (\hat{Y}^n)^{(\tau)} & = \arg\max_{Y^n} p(y(L^n) = Y^n | G^1, \cdots, G^n, (\hat{Y}^1)^{(\tau)}, (\hat{Y}^2)^{(\tau)}, \cdots, (\hat{Y}^{(n-1)})^{(\tau)}) \end{cases} \tag{6.12}$$

The structure of the link prediction framework is shown in Fig. 6.4a. When predicting social links in network G^i, we can extract features based on the intra-network social meta path extracted from G^i and those extracted based on the inter-network social meta path across $G^1, G^2, \cdots, G^{i-1}, G^{i+1}, \cdots, G^n$ for links in P^i, U^i and L^i. Feature vectors $\mathbf{x}(P)$ and $\mathbf{x}(P)$ as well as the labels, $y(P)$, $y(U)$, of links in P and U are passed to the PU link prediction model M^i and the meta path selection model MS^i. The formation probabilities of links in L^i predicted by model M^i will be used to update the network by replace the weights of L^i with the newly predicted formation probabilities. The initial weights of these potential links in L^i are set as 0 (i.e., the formation probability of links mentioned in Definition 11). After finishing these steps on G^i, we will move to conduct similar operations on G^{i+1}. We iteratively predict links in G^1 to G^n alternatively in a sequence until the results in all of these networks converge.

6.2.3 Experiments

To test the effectiveness of the proposed MLI framework, in this section, extensive experiments have been done on two real-world partially co-aligned online social networks dataset introduced in the previous section.

6.2.3.1 Performance Evaluation Results

To show the advantages of MLI, we compare MLI with many other baseline methods, which include:

- MLI: Method MLI is the multinetwork link prediction framework, which can predict social links in multiple online social networks simultaneously. The features used by MLI are extracted based on the meta paths selected from Φ and Ψ across aligned networks.
- LI: Method LI (Link Identifier) is identical to MLI except that LI predict the formation of social links in each network independently.
- SCAN: Method SCAN (Cross Aligned Network link prediction) proposed in [42, 43] is similar to MLI except that (1) SCAN predicts social links in each network independently; (2) features used by SCAN are those extracted based on meta paths Φ and Ψ_1 without meta path selection.
- SCAN- S: Method SCAN- S (SCAN with Source Network) proposed in [42, 43] is identical to SCAN except that the features used by SCAN- S are those extracted based on Ψ_1 without meta path selection.
- SCAN- T: Method SCAN- T (SCAN with Target Network) proposed in [42, 43]) is identical to SCAN except that the features used by SCAN- S are those extracted based on Φ without meta path selection.

The social links in both Foursquare and Twitter are used as the ground truth to evaluate the prediction results. SVM [4] with linear kernel and optimal parameters is used as the base classifier of all comparison methods. Accuracy, AUC, and F1 score are used as the evaluation metrics in the experiments.

To denote different degrees of network newness, in Table 6.5, we fix ρ^T as 0.8 but changes ρ^F within $\{0.1, 0.2, \cdots, 0.8\}$. Table 6.5 has two parts: the upper part is the link prediction results in Foursquare and the lower part is that in Twitter, as MLI is an integrated PU link prediction framework. The link prediction results in each part are evaluated by different metrics: AUC, Accuracy, and F1. As shown in Table 6.5, MLI can outperform all other comparison methods consistently for $\rho^F \in \{0.1, 0.2, \cdots, 0.8\}$ in both Foursquare network and Twitter network. For

Table 6.5 Performance comparison of different methods for inferring social links for Foursquare and Twitter of different remaining information rates. The anchor link sample rate ρ^A is set as 1.0

Network	Measure	Methods	Remaining information rates ρ^F of Foursquare							
			0.1	0.2	0.3	0.4	0.5	0.6	0.7	0.8
Foursquare	AUC	MLI	**0.677±0.023**	**0.776±0.011**	**0.844±0.008**	**0.887±0.005**	**0.906±0.003**	**0.912±0.005**	**0.912±0.003**	**0.916±0.004**
		LI	0.573±0.019	0.68±0.023	0.806±0.01	0.853±0.004	0.866±0.003	0.874±0.007	0.881±0.003	0.878±0.005
		SCAN	0.549±0.009	0.56±0.009	0.662±0.03	0.745±0.009	0.786±0.014	0.804±0.01	0.812±0.005	0.82±0.004
		SCANт	0.5±0.083	0.503±0.007	0.613±0.012	0.739±0.008	0.764±0.013	0.787±0.007	0.8±0.006	0.81±0.007
		SCANs	0.524±0.013	0.524±0.017	0.524±0.012	0.524±0.005	0.524±0.002	0.524±0.01	0.524±0.003	0.524±0.005
	Accuracy	MLI	**0.632±0.01**	**0.692±0.007**	**0.755±0.005**	**0.769±0.004**	**0.779±0.002**	**0.798±0.006**	**0.799±0.004**	**0.797±0.005**
		LI	0.568±0.013	0.624±0.053	0.699±0.004	0.722±0.006	0.761±0.01	0.782±0.01	0.789±0.005	0.791±0.006
		SCAN	0.558±0.007	0.6±0.006	0.683±0.071	0.714±0.009	0.721±0.007	0.736±0.007	0.75±0.008	0.765±0.009
		SCANт	0.491±0.019	0.568±0.004	0.65±0.008	0.685±0.007	0.714±0.007	0.727±0.009	0.736±0.012	0.747±0.003
		SCANs	0.548±0.011	0.548±0.055	0.548±0.007	0.548±0.008	0.548±0.007	0.548±0.01	0.548±0.003	0.548±0.006
	F1	MLI	**0.644±0.01**	**0.695±0.022**	**0.722±0.013**	**0.742±0.005**	**0.761±0.005**	**0.789±0.006**	**0.783±0.005**	**0.786±0.006**
		LI	0.63±0.017	0.635±0.015	0.66±0.007	0.684±0.01	0.715±0.016	0.753±0.014	0.764±0.007	0.766±0.009
		SCAN	0.6±0.02	0.609±0.006	0.614±0.031	0.632±0.018	0.645±0.018	0.676±0.016	0.701±0.01	0.726±0.013
		SCANт	0.534±0.196	0.559±0.004	0.565±0.016	0.584±0.011	0.645±0.011	0.674±0.016	0.696±0.019	0.712±0.01
		SCANs	0.56±0.016	0.56±0.041	0.56±0.015	0.56±0.015	0.56±0.013	0.56±0.013	0.56±0.005	0.56±0.01

(continued)

Table 6.5 (continued)

Network	Measure	Methods	Remaining information rates ρ^F of Foursquare							
			0.1	0.2	0.3	0.4	0.5	0.6	0.7	0.8
Twitter	AUC	MLI	**0.884±0.004**	**0.891±0.003**	**0.915±0.003**	**0.917±0.003**	**0.923±0.002**	**0.929±0.003**	**0.927±0.003**	**0.937±0.003**
		LI	0.841±0.003	0.847±0.002	0.852±0.003	0.862±0.002	0.873±0.002	0.884±0.003	0.894±0.003	0.904±0.003
		SCAN	0.801±0.003	0.814±0.002	0.819±0.003	0.817±0.002	0.819±0.002	0.823±0.003	0.831±0.002	0.837±0.003
		SCANт	0.802±0.002	0.802±0.002	0.802±0.002	0.802±0.002	0.802±0.002	0.802±0.002	0.802±0.002	0.802±0.002
		SCANs	0.508±0.002	0.543±0.002	0.584±0.003	0.631±0.001	0.653±0.002	0.666±0.003	0.673±0.003	0.686±0.003
	Accuracy	MLI	**0.92±0.003**	**0.927±0.002**	**0.927±0.003**	**0.929±0.004**	**0.93±0.003**	**0.932±0.003**	**0.936±0.003**	**0.936±0.004**
		LI	0.899±0.004	0.904±0.004	0.908±0.004	0.913±0.002	0.916±0.003	0.918±0.003	0.918±0.003	0.92±0.004
		SCAN	0.831±0.005	0.835±0.003	0.837±0.006	0.842±0.001	0.844±0.002	0.848±0.004	0.848±0.002	0.849±0.004
		SCANт	0.827±0.003	0.827±0.003	0.827±0.003	0.827±0.003	0.827±0.003	0.827±0.003	0.827±0.003	0.827±0.003
		SCANs	0.568±0.004	0.577±0.003	0.585±0.002	0.587±0.002	0.591±0.003	0.594±0.003	0.596±0.003	0.598±0.004
	F1	MLI	**0.804±0.002**	**0.808±0.002**	**0.809±0.003**	**0.811±0.003**	**0.812±0.003**	**0.818±0.003**	**0.826±0.003**	**0.826±0.004**
		LI	0.776±0.005	0.785±0.005	0.792±0.005	0.8±0.003	0.804±0.003	0.808±0.003	0.809±0.003	0.811±0.004
		SCAN	0.682±0.006	0.686±0.004	0.69±0.006	0.699±0.001	0.703±0.003	0.707±0.004	0.709±0.002	0.711±0.005
		SCANт	0.683±0.003	0.683±0.003	0.683±0.003	0.683±0.003	0.683±0.003	0.683±0.003	0.683±0.003	0.683±0.003
		SCANs	0.53±0.006	0.546±0.006	0.559±0.004	0.564±0.004	0.571±0.004	0.575±0.004	0.581±0.004	0.583±0.005

example, in Foursquare when $\rho^F = 0.5$, the AUC achieved by MLI is about 5% better than LI, 15% better than SCAN, 19% better than SCAN- T and 73% better than SCAN- S; the Accuracy achieved by MLI is about 2.3% better than LI, 8% better than SCAN, 9.1% higher than SCAN- T and over 40% higher than SCAN- S; the F1 of MLI is 6.4% higher than LI, 18% higher than SCAN and SCAN- T and 36% higher than SCAN- S. When $\rho^F = 0.5$, the link prediction results of MLI in Twitter are also much better than all other baseline methods. For instances, in Twitter the AUC of MLI is 0.923 ± 0.002, which is about 6% better than LI, over 13% better than SCAN, SCAN- T and over 40% better than SCAN- S. Similar results can be obtained when evaluated by Accuracy and F1.

In Table 6.6, we fix $\rho^F = 0.8$ but change ρ^T with values in $\{0.1, 0.2, \cdots, 0.8\}$. Similar to the results obtained in Table 6.5 where ρ^F varies, MLI can beat all other methods in both Twitter and Foursquare when the degree of newness of the Twitter network changes.

MLI can perform better than LI in both Foursquare and Twitter, which shows that predicting social links in multiple networks simultaneously in MLI framework can do enhance the results in both networks; the fact that LI can beat SCAN shows that features extracted based on cross network meta paths can do transfer useful information for both anchor and non-anchor users; SCAN works better than both SCAN- T and SCAN- S denotes that link prediction with information in two networks simultaneously is better than that with information in one single network.

6.2.3.2 Parameter Analysis

An important parameter that can affect the performance of all these methods is the rate of anchor links existing across networks. In this part, we will analyze the effects of the anchor link rate, $\rho^A \in [0, 1.0]$. To exclude other parameters' interference, we fix ρ^F and ρ^T as 0.8 but change ρ^A with values in $\{0.1, 0.2, \cdots, 1.0\}$ and study the link prediction results in both Foursquare and Twitter under the evaluation of AUC, Accuracy, and F1. The results are shown in Fig. 6.5.

As shown in Fig. 6.5, where Fig. 6.5a–c are the link prediction results in Foursquare and the Fig. 6.5d–f are those in Twitter, almost all the methods can perform better as ρ^A increases, except SCAN- T as it only utilizes information in the target network only. It shows that with more anchor links, MLI, LI, SCAN and SCAN- S can transfer much more information from other aligned source networks to the target network to enhance the results. In addition, MLI can work better than LI consistently as ρ^A varies, which can show the effectiveness of MLI in dealing with networks with different ratios of anchor links

6.2.3.3 Convergence Analysis

MLI need to predict the links in all the aligned networks alternatively and iteratively until convergence. In this part, we will analyze whether MLI can converge as this

Table 6.6 Performance comparison of different methods for inferring social links for Foursquare and Twitter of different remaining information rates. The anchor link sample rate ρ^A is set as 1.0

Network	Measure	Methods	Remaining information rates ρ^T of Twitter							
			0.1	0.2	0.3	0.4	0.5	0.6	0.7	0.8
Foursquare	AUC	MLI	**0.862±0.003**	**0.867±0.004**	**0.87±0.003**	**0.873±0.005**	**0.885±0.003**	**0.891±0.003**	**0.895±0.004**	**0.916±0.004**
		LI	0.831±0.005	0.834±0.004	0.846±0.004	0.853±0.005	0.855±0.005	0.867±0.004	0.868±0.005	0.87±0.005
		SCAN	0.81±0.007	0.81±0.008	0.812±0.005	0.817±0.007	0.816±0.01	0.815±0.007	0.822±0.006	0.82±0.004
		SCANт	0.81±0.007	0.81±0.007	0.81±0.007	0.81±0.007	0.809±0.007	0.809±0.007	0.81±0.007	0.81±0.007
		SCANs	0.504±0.007	0.51±0.003	0.511±0.005	0.516±0.005	0.522±0.004	0.53±0.005	0.53±0.004	0.53±0.005
	Accuracy	MLI	**0.78±0.003**	**0.786±0.005**	**0.789±0.004**	**0.794±0.005**	**0.793±0.004**	**0.789±0.004**	**0.796±0.005**	**0.797±0.005**
		LI	0.745±0.011	0.762±0.005	0.768±0.007	0.772±0.007	0.777±0.008	0.783±0.008	0.789±0.006	0.791±0.006
		SCAN	0.749±0.007	0.754±0.006	0.754±0.007	0.757±0.006	0.758±0.007	0.761±0.008	0.763±0.009	0.765±0.009
		SCANт	0.748±0.003	0.748±0.003	0.747±0.003	0.748±0.003	0.748±0.003	0.748±0.003	0.748±0.003	0.747±0.003
		SCANs	0.692±0.011	0.717±0.008	0.725±0.008	0.746±0.008	0.741±0.006	0.746±0.006	0.75±0.007	0.758±0.006
	F1	MLI	**0.768±0.004**	**0.774±0.005**	**0.778±0.006**	**0.784±0.006**	**0.785±0.005**	**0.777±0.004**	**0.785±0.006**	**0.786±0.006**
		LI	0.721±0.02	0.734±0.01	0.734±0.012	0.736±0.012	0.744±0.012	0.755±0.011	0.764±0.01	0.766±0.009
		SCAN	0.717±0.01	0.718±0.007	0.714±0.009	0.715±0.009	0.718±0.011	0.72±0.012	0.721±0.013	0.726±0.013
		SCANт	0.713±0.01	0.712±0.01	0.712±0.01	0.713±0.01	0.713±0.01	0.712±0.01	0.713±0.01	0.712±0.01
		SCANs	0.509±0.02	0.514±0.014	0.524±0.014	0.529±0.013	0.54±0.009	0.542±0.007	0.559±0.012	0.559±0.01
Twitter	AUC	MLI	**0.837±0.004**	**0.858±0.004**	**0.905±0.005**	**0.926±0.003**	**0.924±0.002**	**0.932±0.003**	**0.934±0.002**	**0.937±0.003**
		LI	0.772±0.009	0.829±0.008	0.871±0.009	0.887±0.002	0.887±0.002	0.897±0.003	0.899±0.003	0.904±0.003
		SCAN	0.706±0.008	0.771±0.012	0.799±0.009	0.817±0.002	0.819±0.002	0.829±0.003	0.83±0.003	0.834±0.003
		SCANт	0.555±0.133	0.678±0.006	0.753±0.044	0.754±0.019	0.764±0.014	0.781±0.014	0.794±0.003	0.802±0.002
		SCANs	0.687±0.008	0.687±0.002	0.687±0.005	0.687±0.002	0.687±0.002	0.687±0.002	0.687±0.003	0.687±0.003

(continued)

Table 6.6 (continued)

Network	Measure	Methods	Remaining information rates ρ^T of Twitter							
			0.1	0.2	0.3	0.4	0.5	0.6	0.7	0.8
	Accuracy	MLI	**0.821±0.005**	**0.864±0.001**	**0.892±0.008**	**0.914±0.004**	**0.925±0.002**	**0.926±0.004**	**0.936±0.002**	**0.936±0.004**
		LI	0.706±0.002	0.834±0.011	0.877±0.003	0.898±0.005	0.912±0.001	0.92±0.004	0.924±0.002	0.92±0.004
		SCAN	0.594±0.006	0.716±0.009	0.781±0.005	0.801±0.003	0.823±0.002	0.831±0.004	0.842±0.002	0.849±0.004
		SCANt	0.547±0.062	0.645±0.038	0.723±0.048	0.786±0.004	0.8±0.002	0.815±0.005	0.824±0.002	0.827±0.003
		SCANs	0.59±0.009	0.59±0.007	0.59±0.004	0.59±0.004	0.59±0.002	0.59±0.004	0.59±0.003	0.59±0.004
	F1	MLI	**0.713±0.009**	**0.762±0.005**	**0.791±0.006**	**0.81±0.004**	**0.81±0.002**	**0.819±0.004**	**0.821±0.002**	**0.826±0.004**
		LI	0.651±0.006	0.671±0.023	0.749±0.014	0.779±0.007	0.801±0.003	0.813±0.005	0.818±0.003	0.811±0.004
		SCAN	0.6±0.017	0.633±0.023	0.657±0.013	0.684±0.004	0.703±0.004	0.714±0.005	0.716±0.002	0.711±0.005
		SCANt	0.552±0.113	0.574±0.016	0.604±0.031	0.618±0.003	0.63±0.001	0.641±0.004	0.67±0.002	0.686±0.003
		SCANs	0.575±0.025	0.575±0.016	0.575±0.005	0.575±0.006	0.575±0.004	0.575±0.004	0.575±0.003	0.575±0.005

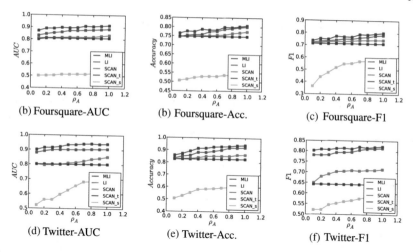

Fig. 6.5 Effects of anchor link ratio ρ^A on prediction results in different networks evaluated by different metrics

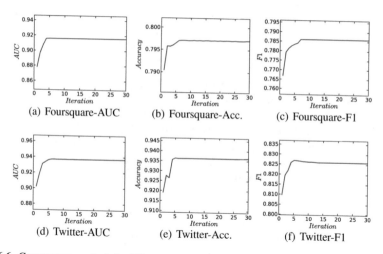

Fig. 6.6 Convergence analysis in different networks under the evaluation of different metrics

process continues. We show the link prediction results achieved by MLI in both Foursquare and Twitter under the evaluation of AUC, Accuracy and F1 when ρ^F, ρ^T and ρ^A are all set as 0.8 in Fig. 6.6. Figure 6.6a–c are the results in Foursquare network from iteration 1 to iteration 30 and Fig. 6.6d–f are those in Twitter network. As shown in these figures, results achieved by MLI can converge in less than 10 iterations in both Foursquare and Twitter evaluated by all these three metrics.

6.3 Synergistic Network Community Detection

6.3.1 Overview

Clustering is a very broad research area, which includes various types of clustering problems, e.g., consensus clustering [20, 21], multiview clustering [2, 3], multirelational clustering [35], co-training-based clustering [15], at the same time. Clustering-based community detection in online social networks is a hot research topic and many different models have already been proposed to optimizing certain evaluation metrics, e.g., modularity function [25] and normalized cut [29]. A detailed survey about existing community detection works is available in [23, 24]. Meanwhile, based on the information available in multiple aligned networks, Jin [9], Zhang et al. [36] and Shao et al. [28] propose to do synergistic community detection across multiple aligned social networks. Via the anchor links, Zhang et al. also propose to transfer information from developed networks to detect social community structures in emerging networks in [48].

The goal of cross-network community detection is to distill relevant information from another social network to compliment knowledge directly derivable from each network to improve the clustering or community detection, while preserving the distinct characteristics of each individual network. To solve the mutual clustering problem, a novel community detection method, MCD, is proposed in [36]. By mapping the social network relations into a heterogeneous information, the proposed method in [36] uses the concept of social meta path to define closeness measure among users. Based on this similarity measure, the proposed method [36] can preserve the network characteristics and utilize the information in other networks to refine community structures mutually at the same time. In this section, we will introduce the mutual community detection framework proposed in [36] briefly.

6.3.2 Cross-Network Community Detection

Given multiple aligned networks $G = (\{G^1, G^2, \cdots, G^n\}, \{A^{(1,2)}, A^{(1,3)}, \cdots, A^{(n-1,n)}\})$, the cross-network community detection problem aims at detecting the community structures of networks $G^1, G^2, \cdots, G^n$, respectively.

6.3.2.1 Network Characteristic Preservation Clustering

Clustering each network independently can preserve each networks characteristics effectively as no information from external networks will interfere with the clustering results. Partitioning users of a certain network into several clusters will cut connections in the network and lead to some costs inevitably. Optimal clustering results can be achieved by minimizing the clustering costs.

Let $\mathbf{A}_i$ be the adjacency matrix corresponding to the intra-network meta path # i among users in the network and $\mathbf{A}_i(m, n) = k$ iff there exist k different path instances of intra-network meta path # i from user m to n in the network. Furthermore, the similarity score matrix among users of meta path # i can be represented as $\mathbf{S}_i = \left(\mathbf{D}_i + \bar{\mathbf{D}}_i\right)^{-1}\left(\mathbf{A}_i + \mathbf{A}_i^T\right)$, where $\mathbf{A}_i^T$ denotes the transpose of $\mathbf{A}_i$, diagonal matrices $\mathbf{D}_i$ and $\bar{\mathbf{D}}_i$ have values $\mathbf{D}_i(l, l) = \sum_m \mathbf{A}_i(l, m)$ and $\bar{\mathbf{D}}_i(l, l) = \sum_m (\mathbf{A}_i^T)(l, m)$ on their diagonals, respectively. The meta path-based similarity matrix of the network which can capture all possible connections among users is represented as follows:

$$\mathbf{S} = \sum_i \omega_i \mathbf{S}_i = \sum_i \omega_i \left(\left(\mathbf{D}_i + \bar{\mathbf{D}}_i\right)^{-1}\left(\mathbf{A}_i + \mathbf{A}_i^T\right)\right). \qquad (6.13)$$

For a given network G, let $C = \{U_1, U_2, \ldots, U_k\}$ be the community structures detected from G. Term $\overline{U}_i = U - U_i$ is defined to be the complement of set U_i in G. Various cost measure of partition C can be used, e.g., cut and normalized cut:

$$cut(C) = \frac{1}{2} \sum_{i=1}^{k} S(U_i, \overline{U}_i) = \frac{1}{2} \sum_{i=1}^{k} \sum_{u \in U_i, v \in \overline{U}_i} S(u, v), \qquad (6.14)$$

$$Ncut(C) = \frac{1}{2} \sum_{i=1}^{k} \frac{S(U_i, \overline{U}_i)}{S(U_i, \cdot)} = \sum_{i=1}^{k} \frac{cut(U_i, \overline{U}_i)}{S(U_i, \cdot)}, \qquad (6.15)$$

where $S(u, v)$ denotes the similarity between u, v and $S(U_i, \cdot) = S(U_i, U) = S(U_i, U_i) + S(U_i, \overline{U}_i)$.

For all users in U, their clustering result can be represented in the result confidence matrix $\mathbf{H}$, where $\mathbf{H} = [\mathbf{h}_1, \mathbf{h}_2, \ldots, \mathbf{h}_n]^T$, $n = |U|$, $\mathbf{h}_i = (h_{i,1}, h_{i,2}, \ldots, h_{i,k})$ and $h_{i,j}$ denotes the confidence that $u_i \in U$ is in cluster $U_j \in C$. The optimal $\mathbf{H}$ that can minimize the normalized-cut cost can be obtained by solving the following objective function:

$$\min_{\mathbf{H}} \text{Tr}(\mathbf{H}^T \mathbf{L} \mathbf{H}),$$

$$s.t. \mathbf{H}^T \mathbf{D} \mathbf{H} = \mathbf{I}. \qquad (6.16)$$

where $\mathbf{L} = \mathbf{D} - \mathbf{S}$, diagonal matrix $\mathbf{D}$ has $\mathbf{D}(i, i) = \sum_j S(i, j)$ on its diagonal, and $\mathbf{I}$ is an identity matrix.

6.3.2.2 Clustering of Multiple Aligned Networks

Besides the shared information due to common network construction purposes and similar network features [48], anchor users can also have unique information (e.g., social structures) across aligned networks, which can provide us with a more comprehensive knowledge about the community structures formed by these users. Meanwhile, by maximizing the consensus (i.e., minimizing the "discrepancy") of the clustering results about the anchor users in multiple partially aligned networks, we refine the clustering results of the anchor users with information in other aligned networks mutually. We can represent the clustering results achieved in G^i and G^j as $C^i = \{U_1^i, U_2^i, \cdots, U_{k^i}^i\}$ and $C^j = \{U_1^j, U_2^j, \cdots, U_{k^j}^j\}$, respectively.

Let u_p and u_q be two anchor users in the network, whose accounts in G^i and G^j are u_p^i, u_p^j, u_q^i and u_q^j, respectively. If users u_p^i and u_q^i are partitioned into the same cluster in G^i but their corresponding accounts u_p^j and u_q^j are partitioned into different clusters in G^j, then it will lead to a discrepancy between the clustering results of u_p^i, u_p^j, u_q^i and u_q^j in aligned networks G^i and G^j.

Definition 6.10 (*Discrepancy*) The discrepancy between the clustering results of u_p and u_q across aligned networks G^i and G^j is defined as the difference of confidence scores of u_p and u_q being partitioned in the same cluster across aligned networks. Considering that in the clustering results, the confidence scores of u_p^i and u_q^i (u_p^j and u_q^j) being partitioned into k^i (k^j) clusters can be represented as vectors $\mathbf{h}_p^i$ and $\mathbf{h}_q^i$ ($\mathbf{h}_p^j$ and $\mathbf{h}_q^j$), respectively, while the confidences that u_p and u_q are in the same cluster in G^i and G^j can be denoted as $\mathbf{h}_p^i (\mathbf{h}_q^i)^T$ and $\mathbf{h}_p^j (\mathbf{h}_q^j)^T$. Formally, the discrepancy of the clustering results about u_p and u_q is defined to be $d_{p,q}(C^i, C^j) = \left(\mathbf{h}_p^i (\mathbf{h}_q^i)^T - \mathbf{h}_p^j (\mathbf{h}_q^j)^T \right)^2$ if u_p, u_q are both anchor users; and $d_{p,q}(C^i, C^j) = 0$ otherwise. Furthermore, the discrepancy of C^i and C^j will be:

$$d(C^i, C^j) = \sum_p^{n^i} \sum_q^{n^j} d_{p,q}(C^i, C^j), \qquad (6.17)$$

where $n^i = |U^i|$ and $n^j = |U^j|$.

However, considering that $d(C^i, C^j)$ is highly dependent on the number of anchor users and anchor links between G^i and G^j, minimizing $d(C^i, C^j)$ can favor highly consented clustering results when the anchor users are abundant but have no significant effects when the anchor users are very rare. To solve this problem, we propose to minimize the normalized discrepancy instead.

Definition 6.11 (*Normalized Discrepancy*) The normalized discrepancy measure computes the differences of clustering results in two aligned networks as a fraction of the discrepancy with regard to the number of anchor users across partially aligned networks:

$$Nd(C^i, C^j) = \frac{d(C^i, C^j)}{\left(|A^{(i,j)}|\right)\left(|A^{(i,j)}| - 1\right)}. \qquad (6.18)$$

Optimal consensus clustering results of G^i and G^j will be $\hat{\mathrm{C}}^i, \hat{\mathrm{C}}^j$:

$$\hat{\mathrm{C}}^i, \hat{\mathrm{C}}^j = \arg \min_{\mathrm{C}^i, \mathrm{C}^j} Nd(\mathrm{C}^i, \mathrm{C}^j). \qquad (6.19)$$

Similarly, the normalized-discrepancy objective function can also be represented with the clustering results confidence matrices $\mathbf{H}^i$ and $\mathbf{H}^j$ as well. Meanwhile, considering that the networks studied in this chapter are partially aligned, matrices $\mathbf{H}^i$ and $\mathbf{H}^j$ contain the results of both anchor users and non-anchor users, while non-anchor users should not be involved in the discrepancy calculation according to the definition of discrepancy. After pruning the non-anchor users from the confidence matrices, we can represent the pruned confidence matrices as $\bar{\mathbf{H}}^i$ and $\bar{\mathbf{H}}^j$.

Furthermore, the objective function of inferring clustering confidence matrices, which can minimize the normalized discrepancy can be represented as follows

$$\min_{\mathbf{H}^i, \mathbf{H}^j} \frac{\left\| \bar{\mathbf{H}}^i \left(\bar{\mathbf{H}}^i\right)^T - \bar{\mathbf{H}}^j \left(\bar{\mathbf{H}}^j\right)^T \right\|_F^2}{\left\| \mathbf{T}^{(i,j)} \right\|_F^2 \left(\left\| \mathbf{T}^{(i,j)} \right\|_F^2 - 1 \right)},$$
$$s.t. (\mathbf{H}^i)^T \mathbf{D}^i \mathbf{H}^i = \mathbf{I}, (\mathbf{H}^j)^T \mathbf{D}^j \mathbf{H}^j = \mathbf{I}. \qquad (6.20)$$

where $\mathbf{D}^i, \mathbf{D}^j$ are the corresponding diagonal matrices of similarity matrices of networks G^i and G^j, respectively.

6.3.2.3 Joint Optimization Objective Function

Taking both of these two issues into considerations, the optimal mutual clustering results $\hat{\mathrm{C}}^i$ and $\hat{\mathrm{C}}^j$ of aligned networks G^i and G^j can be achieved as follows:

$$\arg \min_{\mathrm{C}^i, \mathrm{C}^j} \alpha \cdot Ncut(\mathrm{C}^i) + \beta \cdot Ncut(\mathrm{C}^j) + \theta \cdot Nd(\mathrm{C}^i, \mathrm{C}^j) \qquad (6.21)$$

where $\alpha, \beta,$ and θ represent the weights of these terms and, for simplicity, α and β are both set as 1.

By replacing $Ncut(\mathrm{C}^i)$, $Ncut(\mathrm{C}^j)$, $Nd(\mathrm{C}^i, \mathrm{C}^j)$ with the objective equations derived above, we can rewrite the joint objective function as follows:

$$\min_{\mathbf{H}^i, \mathbf{H}^j} \alpha \cdot \mathrm{Tr}((\mathbf{H}^i)^T \mathbf{L}^i \mathbf{H}^i) + \beta \cdot \mathrm{Tr}((\mathbf{H}^j)^T \mathbf{L}^j \mathbf{H}^j) + \theta \cdot \frac{\left\| \bar{\mathbf{H}}^i \left(\bar{\mathbf{H}}^i\right)^T - \bar{\mathbf{H}}^j \left(\bar{\mathbf{H}}^j\right)^T \right\|_F^2}{\left\| \mathbf{T}^{(i,j)} \right\|_F^2 \left(\left\| \mathbf{T}^{(i,j)} \right\|_F^2 - 1 \right)},$$
$$s.t. (\mathbf{H}^i)^T \mathbf{D}^i \mathbf{H}^i = \mathbf{I}, (\mathbf{H}^j)^T \mathbf{D}^j \mathbf{H}^j = \mathbf{I}, \qquad (6.22)$$

where $\mathbf{L}^i = \mathbf{D}^i - \mathbf{S}^i, \mathbf{L}^j = \mathbf{D}^j - \mathbf{S}^j$ and matrices $\mathbf{S}^i, \mathbf{S}^j$ and $\mathbf{D}^i, \mathbf{D}^j$ are the similarity matrices and their corresponding diagonal matrices defined before.

The objective function is a complex optimization problem with orthogonality constraints, which can be very difficult to solve because the constraints are not only non-convex, but also numerically expensive to preserve during iterations. Please refer to [36] for more information about the solution to the objective function.

6.3.3   Experiments

To test the performance of the MCD model in detecting the communities across multiple aligned social networks, extensive experiments have been done on the aligned social networks dataset: Foursquare and Twitter. The experimental results will be illustrated as follows.

6.3.3.1   Performance Evaluation Results

The comparison methods used in the experiments can be divided into three categories, **Mutual Clustering Methods**

- MCD: MCD is the mutual community detection method, which can detect the communities of multiple aligned networks with consideration of the connections and characteristics of different networks. Heterogeneous information in multiple aligned networks are applied in building MCD.

Multinetwork Clustering Methods

- SICLUS: the clustering method proposed in [38, 48] can calculate the similarity scores among users by propagating heterogeneous information across views/networks. We extend the method proposed in [38, 48] and propose SICLUS to calculate the intimacy scores among users in multiple networks simultaneously, based on which, users can be grouped into different clusters with clustering models based on intimacy matrix factorization as introduced in [48]. Heterogeneous information across networks is used to build SICLUS.

Isolated Clustering Methods, which can detect communities in each isolated network:

- NCUT: NCUT is the clustering method based on normalized cut proposed in [29]. Method NCUT can detect the communities in each social network merely based on the social connections in each network in the experiments.
- KMEANS: KMEANS is a traditional clustering method, which can be used to detect communities [27] in social networks based on the social connections only in the experiments.

The evaluation metrics applied can be divided into two categories: Quality Metrics and Consensus Metrics.

Quality Metrics: The four widely and commonly used quality metrics are applied to measure the clustering result, e.g., $C = \{U_i\}_{i=1}^K$, of each network.

- **normalized-dbi** [38]:

$$
ndbi(C) = \frac{1}{K} \sum_i \min_{j \neq i} \frac{d(c_i, c_j) + d(c_j, c_i)}{\sigma_i + \sigma_j + d(c_i, c_j) + d(c_j, c_i)}, \tag{6.23}
$$

where c_i is the centroid of community $U_i \in C$, $d(c_i, c_j)$ denotes the distance between centroids c_i and c_j and σ_i represents the average distance between elements in U_i and centroid c_i. (Higher ndbi corresponds to better performance).
- **entropy** [38]: $H(C) = -\sum_{i=1}^K P(i) \log P(i)$, where $P(i) = \frac{|U_i|}{\sum_{i=1}^K |U_i|}$. (Lower entropy corresponds to better performance).
- **density** [38]: $dens(C) = \sum_{i=1}^K \frac{|E_i|}{|E|}$, where E and E_i are the edge sets in the network and U_i. (Higher density corresponds to better performance).
- **silhouette** [19]:

$$
sil(C) = \frac{1}{K} \sum_{i=1}^K \left(\frac{1}{|U_i|} \sum_{u \in U_i} \frac{b(u) - a(u)}{\max\{a(u), b(u)\}} \right), \tag{6.24}
$$

where $a(u) = \frac{1}{|U_i|-1} \sum_{v \in U_i, u \neq v} d(u, v)$ and $b(u) = \min_{j, j \neq i} \left(\frac{1}{|U_j|} \sum_{v \in U_j} d(u, v) \right)$. (Higher silhouette corresponds to better performance).

Consensus Metrics: Given the clustering results $C^{(1)} = \{U_i^{(1)}\}_{i=1}^{K^{(1)}}$ and $C^{(2)} = \{U_i^{(2)}\}_{i=1}^{K^{(2)}}$, the consensus metrics measuring the how similar or dissimilar the anchor users are clustered in $C^{(1)}$ and $C^{(2)}$ include:

- **rand** [26]: $rand(C^{(1)}, C^{(2)}) = \frac{N_{01} + N_{10}}{N_{00} + N_{01} + N_{10} + N_{11}}$, where $N_{11}(N_{00})$ is the numbers of pairwise anchor users who are clustered in the same (different) community(ies) in both $C^{(1)}$ and $C^{(2)}$, $N_{01}(N_{10})$ is that of anchor users who are clustered in the same community (different communities) in $C^{(1)}$ but in different communities (the same communities) in $C^{(2)}$. (Lower rand corresponds to better performance).
- **variation of information** (vi) [26]: $vi(C^{(1)}, C^{(2)}) = H(C^{(1)}) + H(C^{(2)}) - 2mi(C^{(1)}, C^{(2)})$. (Lower vi corresponds to better performance).
- **mutual information** [26]: $mi(C^{(1)}, C^{(2)}) = \sum_{i=1}^{K^{(1)}} \sum_{j=1}^{K^{(2)}} P(i, j) \log \frac{P(i,j)}{P(i)P(j)}$, where $P(i, j) = \frac{|U_i^{(1)} \cap_A U_j^{(2)}|}{|A|}$ and $|U_i^{(1)} \cap_A U_j^{(2)}| = \left| \{u | u \in U_i^{(1)}, \exists v \in U_j^{(2)}, (u, v) \in A\} \right|$ [12]. (Higher mi corresponds to better performance).
- **normalized mutual information** [26]: $nmi(C^{(1)}, C^{(2)}) = \frac{mi(C^{(1)}, C^{(2)})}{\sqrt{H(C^{(1)})H(C^{(2)})}}$. (Higher nmi corresponds to better performance).

The experiment results are available in Tables 6.7 and 6.8. To show the effects of the anchor links, we use the same networks but randomly sample a proportion of anchor links from the networks, whose number is controlled by $\sigma \in$

Table 6.7 Community detection results of foursquare and twitter evaluated by quality metrics

Network	Measure	Methods	Remaining anchor link rates σ									
			0.1	0.2	0.3	0.4	0.5	0.6	0.7	0.8	0.9	1.0
Foursquare	ndbi	MCD	**0.927**	**0.924**	**0.95**	**0.969**	**0.966**	**0.961**	**0.958**	**0.954**	**0.971**	**0.958**
		SICLUS	0.891	0.889	0.88	0.877	0.894	0.883	0.89	0.88	0.887	0.893
		NCUT	0.863	0.863	0.863	0.863	0.863	0.863	0.863	0.863	0.863	0.863
		KMEANS	0.835	0.835	0.835	0.835	0.835	0.835	0.835	0.835	0.835	0.835
	Entropy	MCD	**1.551**	**1.607**	**1.379**	**1.382**	**1.396**	**1.382**	**1.283**	**1.552**	**1.308**	**1.497**
		SICLUS	4.332	4.356	4.798	4.339	4.474	4.799	4.446	4.658	4.335	4.459
		NCUT	2.768	2.768	2.768	2.768	2.768	2.768	2.768	2.768	2.768	2.768
		KMEANS	2.369	2.369	2.369	2.369	2.369	2.369	2.369	2.369	2.369	2.369
	Density	MCD	**0.216**	**0.205**	**0.196**	0.163	**0.239**	**0.192**	**0.303**	**0.198**	0.170	**0.311**
		SICLUS	0.116	0.121	0.13	0.095	0.143	0.11	0.13	0.12	0.143	0.103
		NCUT	0.154	0.154	0.154	0.154	0.154	0.154	0.154	0.154	0.154	0.154
		KMEANS	0.182	0.182	0.182	**0.182**	0.182	0.182	0.182	0.182	**0.182**	0.182
	Silhouette	MCD	**-0.137**	**-0.114**	**-0.148**	**-0.156**	**-0.117**	**-0.11**	**-0.035**	**-0.125**	**-0.148**	**-0.044**
		SICLUS	-0.168	-0.198	-0.173	-0.189	-0.178	-0.181	-0.21	-0.195	-0.167	-0.18
		NCUT	-0.34	-0.34	-0.34	-0.34	-0.34	-0.34	-0.34	-0.34	-0.34	-0.34
		KMEANS	-0.297	-0.297	-0.297	-0.297	-0.297	-0.297	-0.297	-0.297	-0.297	-0.297
Twitter	ndbi	MCD	**0.962**	**0.969**	**0.955**	**0.969**	**0.97**	**0.958**	**0.952**	**0.96**	**0.946**	**0.953**
		SICLUS	0.815	0.843	0.807	0.83	0.826	0.832	0.835	0.808	0.812	0.836
		NCUT	0.759	0.759	0.759	0.759	0.759	0.759	0.759	0.759	0.759	0.759
		KMEANS	0.761	0.761	0.761	0.761	0.761	0.761	0.761	0.761	0.761	0.761
	Entropy	MCD	**2.27**	**2.667**	**2.48**	**2.381**	**2.43**	**2.372**	**2.452**	**2.459**	**2.564**	**2.191**
		SICLUS	4.780	5.114	5.066	4.961	4.904	4.866	5.121	4.629	4.872	5.000
		NCUT	3.099	3.099	3.099	3.099	3.099	3.099	3.099	3.099	3.099	3.099
		KMEANS	3.245	3.245	3.245	3.245	3.245	3.245	3.245	3.245	3.245	3.245

(continued)

Table 6.7 (continued)

Network	Measure	Methods	Remaining anchor link rates σ									
			0.1	0.2	0.3	0.4	0.5	0.6	0.7	0.8	0.9	1.0
	Density	MCD	**0.14**	0.097	**0.142**	0.109	**0.15**	**0.158**	**0.126**	**0.149**	**0.147**	**0.164**
		SICLUS	0.055	0.017	0.044	0.026	0.04	0.062	0.016	0.044	0.045	0.02
		NCUT	0.107	0.107	0.107	0.107	0.107	0.107	0.107	0.107	0.107	0.107
		KMEANS	0.119	**0.119**	0.119	**0.119**	0.119	0.119	0.119	0.119	0.119	0.119
	Silhouette	MCD	**-0.137**	**-0.179**	**-0.282**	**-0.175**	**-0.275**	**-0.273**	**-0.248**	**-0.269**	**-0.266**	**-0.286**
		SICLUS	-0.356	-0.322	-0.311	-0.347	-0.346	-0.349	-0.323	-0.363	-0.345	-0.352
		NCUT	-0.424	-0.424	-0.424	-0.424	-0.424	-0.424	-0.424	-0.424	-0.424	-0.424
		KMEANS	-0.406	-0.406	-0.406	-0.406	-0.406	-0.406	-0.406	-0.406	-0.406	-0.406

Table 6.8 Community detection results of foursquare and twitter evaluated by consensus metrics

Measure	Methods	Remaining anchor link rates σ									
		0.1	0.2	0.3	0.4	0.5	0.6	0.7	0.8	0.9	1.0
rand	MCD	**0.095**	**0.099**	**0.107**	**0.138**	**0.116**	**0.121**	**0.132**	**0.106**	**0.089**	**0.159**
	SICLUS	0.135	0.139	0.144	0.148	0.142	0.14	0.132	0.132	0.144	**0.141**
	NCUT	0.399	0.377	0.372	0.4	0.416	0.423	0.362	0.385	0.362	0.341
	KMEANS	0.436	0.387	0.4	0.358	0.403	0.363	0.408	0.365	0.35	0.363
vi	MCD	**3.309**	**4.052**	**4.058**	**3.902**	**4.038**	**4.348**	**3.973**	**3.944**	**4.078**	**2.911**
	SICLUS	7.56	8.324	8.414	8.713	8.756	8.836	8.832	8.621	8.427	8.02
	NCUT	5.384	5.268	5.221	4.855	5.145	5.541	5.909	5.32	5.085	5.246
	KMEANS	5.427	5.117	5.355	5.326	5.679	5.944	5.452	5.567	5.513	4.686
nmi	MCD	0.152	**0.152**	**0.149**	**0.141**	**0.149**	**0.156**	**0.142**	**0.158**	**0.147**	0.146
	SICLUS	**0.172**	0.097	0.081	0.06	0.056	0.069	0.078	0.093	0.105	**0.149**
	NCUT	0.075	0.074	0.111	0.108	0.109	0.099	0.05	0.036	0.042	0.106
	KMEANS	0.008	0.047	0.048	0.054	0.048	0.028	0.047	0.014	0.067	0.119
mi	MCD	0.756	**0.611**	**0.4**	0.258	**0.394**	**0.431**	**0.381**	**0.533**	**0.697**	0.689
	SICLUS	**0.780**	0.446	0.367	**0.277**	0.258	0.325	0.374	0.44	0.489	**0.698**
	NCUT	0.188	0.181	0.261	0.232	0.252	0.243	0.138	0.092	0.111	0.31
	KMEANS	0.02	0.112	0.119	0.135	0.127	0.078	0.119	0.038	0.194	0.314

$\{0.1, 0.2, \cdots, 1.0\}$, where $\sigma = 0.1$ means that 10% of all the anchor links are preserved and $\sigma = 1.0$ means that all the anchor links are preserved.

Table 6.7 displays the clustering results of different methods in Foursquare and Twitter, respectively, under the evaluation of ndbi, entropy, density, and silhouette. As shown in these two tables, MCD can achieve the highest ndbi score in both Foursquare and Twitter for different sample rate of anchor links consistently. The entropy of the clustering results achieved by MCD is the lowest among all other comparison methods and is about 70% lower than SICLUS, 40% lower than NCUT and KMEANS in both Foursquare and Twitter. In each community detected by MCD, the social connections are denser than that of SICLUS, NCUT, and KMEANS. Similar results can be obtained under the evaluation of silhouette, the silhouette score achieved by MCD is the highest among all comparison methods. So, MCD can achieve better results than modified multiview and isolated clustering methods under the evaluation of quality metrics.

Table 6.8 shows the clustering results on the aligned networks under the evaluation of consensus metrics, which include rand, vi, nmi, and mi. As shown in Table 6.8, MCD can perform the best among all the comparison methods under the evaluation of consensus metrics. For example, the rand score of MCD is the lowest among all other methods and when $\sigma = 0.5$, the rand score of MCD is 20% lower than SICLUS, 72% lower than NCUT and KMEANS. Similar results can be obtained for other evaluation metrics, like when $\sigma = 0.5$, the vi score of MCD is about half of the score of SICLUS; the nmi and mi score of MCD is the triple of that of KMEANS. As a result, MCD can achieve better performance than both modified multiview and isolated clustering methods under the evaluation of consensus metrics.

According to the results shown in Tables 6.7 and 6.8, we observe that the performance of MCD does not varies much as σ changes. The possible reason can be that, in method MCD, normalized clustering discrepancy is applied to infer the clustering confidence matrices. As σ increases in the experiments, more anchor links are added between networks, part of whose effects will be neutralized by the normalization of clustering discrepancy and does not affect the performance of MCD much.

6.3.3.2 Convergence Analysis

MCD can compute the solution of the optimization function with Curvilinear Search method, which can update matrices $\mathbf{X}^{(1)}$ and $\mathbf{X}^{(2)}$ alternatively. This process will continue until convergence. To check whether this process can stop or not, in this part, we will analyze the convergence of $\mathbf{X}^{(1)}$ and $\mathbf{X}^{(2)}$. In Fig. 6.7, we show the L^1 norm of matrices $\mathbf{X}^{(1)}$ and $\mathbf{X}^{(2)}$, $\left\| \mathbf{X}^{(1)} \right\|_1$ and $\left\| \mathbf{X}^{(2)} \right\|_1$, in each iteration of the updating algorithm, where the L^p norm of matrix $\mathbf{X}$ is $\|\mathbf{X}\|_p = (\sum_i \sum_i X_{ij}^p)^{\frac{1}{p}}$. As shown in Fig. 6.7, both $\left\| \mathbf{X}^{(1)} \right\|_1$ and $\left\| \mathbf{X}^{(2)} \right\|_1$ can converge in less than 200 iterations.

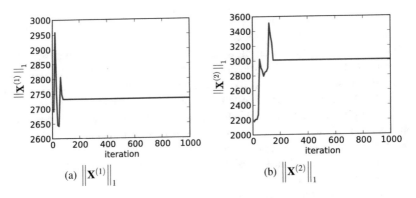

Fig. 6.7 $\left\|\mathbf{X}^{(1)}\right\|_1$ and $\left\|\mathbf{X}^{(2)}\right\|_1$ in each iteration

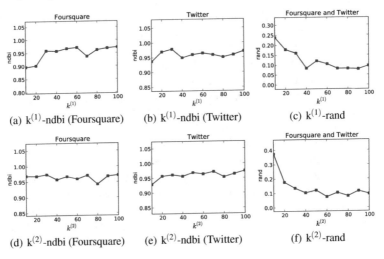

Fig. 6.8 Analysis of parameters $k^{(1)}$ and $k^{(2)}$

6.3.3.3 Parameter Analysis

In method MCD, we have three parameters: $k^{(1)}$, $k^{(2)}$, and θ, where $k^{(1)}$ and $k^{(2)}$ are the numbers of clusters in Foursquare and Twitter networks, respectively, while θ is the weight of the normalized discrepancy term in the object function. In the pervious experiment, we set $k^{(1)} = 50$, $k^{(2)} = 50$ and $\theta = 1.0$. Here, we will analyze the sensitivity of these parameters in details.

To analyze $k^{(1)}$, we fix $k^{(2)} = 50$ and $\theta = 1.0$ but assign $k^{(1)}$ with values in $\{10, 20, 30, 40, 50, 60, 70, 80, 90, 100\}$. The clustering results of MCD with different $k^{(1)}$ evaluated by *ndbi* and *rand* metrics are given in Fig. 6.8a–c. As shown in the figures, the results achieved by MCD are very stable for $k^{(1)}$ with in range [40, 100]

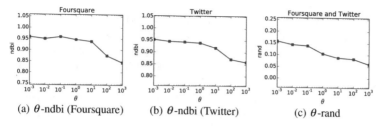

Fig. 6.9 Analysis of parameter θ

under the evaluation of *ndbi* in both Foursquare and Twitter. Similar results can be obtained in Fig. 6.8c, where the performance of MCD on aligned networks is not sensitive to the choice of $k^{(1)}$ for $k^{(1)}$ in range [40, 100] under the evaluation of both *rand*. In a similar way, we can study the sensitivity of parameter $k^{(2)}$, the results about which are shown in Fig. 6.8d–f.

To analyze the parameter θ, we set both $k^{(1)}$ and $k^{(2)}$ as 50 but assign θ with values in $\{0.001, 0.01, 0.1, 1.0, 10.0, 100.0, 1000.0\}$. The results are shown in Fig. 6.9, where when θ is small, e.g., 0.001, the *ndbi* scores achieved by MCD in both Foursquare and Twitter are high but the *rand* score is not good (*rand* is inversely proportional). On the other hand, large θ can lead to good *rand* score but bad *ndbi* scores in both Foursquare and Twitter. As a result, (1) large θ prefers consensus results, (2) small θ can preserve network characteristics and prefers high quality results.

6.4 Conclusions

In this chapter, we have introduced several research works across multiple aligned social networks, including the network alignment problem, link transfer problem, and community detection problem. The problems introduced in this chapter are all very important for many concrete real-world social network applications and services. Several nontrivial algorithms have been proposed to resolve these problems, respectively, whose performance are evaluated with several real-world datasets.

Besides the works introduced in this chapters, many other research problems have been studied across the aligned social networks, like network embedding, information diffusion, viral marketing, and tipping user detection. There are also several interesting directions for further research in the domain of social network fusion learning studies:

- **Multiple Aligned Social Sites**: Existing aligned network studies mainly focus on studying two aligned networks. Meanwhile, when it comes to multiple aligned networks (more than two), many of the studied problems will encounter many new challenges, e.g., the balance of information from different sites, constraints introduced by the multiple sources (e.g., on anchor links).

- **Large Scale Networks**: Most of the introduced methods and models work very well for small-sized social networks, but when it comes to the large scale networks they will suffer from the high time complexity problem a lot. Extending and generalize the existing models to the scalable version will be an interesting direction.
- **Domain Difference Problem**: Many of the existing cross-network studies tackle the domain difference problem in a very simple way, e.g., the meta path selection in link prediction, and meta path weighting in community detection and information diffusion. A more general and effective method to handle the domain difference problem is still an open problem so far.

References

1. Al Hasan, M., Chaoji, V., Salem, S., Zaki, M.: Link prediction using supervised learning. In: SDM (2006)
2. Bickel, S., Scheffer, T.: Multi-view clustering. In: ICDM, pp. 19–26 (2004)
3. Cai, X., Nie, F., Huang, H.: Multi-view k-means clustering on big data. In: IJCAI, pp. 2598–2604 (2013)
4. Chang, C.C., Lin, C.J.: Libsvm: a library for support vector machines. ACM Trans. Intell. Syst. Technol. **2**(3), 389–396 (2011)
5. Dong, Y., Tang, J., Wu, S., Tian, J.: Link prediction and recommendation across heterogeneous social networks. In: ICDM, pp. 181–190 (2012)
6. Elkan, C., Noto, K.: Learning classifiers from only positive and unlabeled data. In: KDD, pp. 213–220 (2008)
7. Getoor, L., Diehl, C.P.: Link mining: a survey. ACM Sigkdd Explor. Newsl. **7**(2), 3–12 (2005)
8. Hasan, M.A., Zaki, M.J.: A survey of link prediction in social networks. In: Aggarwal, C.C. (ed.) Social Network Data Analytics, pp. 243–275. Springer, Berlin (2011)
9. Jin, S., Zhang, J., Yu, P.S., Yang, S.: Synergistic partitioning in multiple large scale social networks. In: IEEE BigData, pp. 281–290 (2014)
10. Klau, G.W.: A new graph-based method for pairwise global network alignment. BMC Bioinform. **10**(1), 135–135 (2009)
11. Kong, X., Yu, P.S., Ding, Y., Wild, D.J.: Meta path-based collective classification in heterogeneous information networks. In: CIKM, pp. 1567–1571 (2012)
12. Kong, X., Zhang, J., Yu, P.S.: Inferring anchor links across multiple heterogeneous social networks. In: CIKM, pp. 179–188 (2013)
13. Koutra, D., Tong, H., Lubensky, D.: Big-align: Fast bipartite graph alignment. In: ICDM, pp. 389–398 (2013)
14. Kuchaiev, O., Milenković, T., Memišević, V., Hayes, W., Pržulj, N.: Topological network alignment uncovers biological function and phylogeny. J. R. Soc. Interface **7**(50), 1341–1354 (2009)
15. Kumar, A.: A co-training approach for multi-view spectral clustering. In: ICML, pp. 393–400 (2011)
16. Li, Y., Shi, C., Philip, S.Y., Chen, Q.: Hrank: a path based ranking method in heterogeneous information network. In: Web-Age Information Management, pp. 553–565 (2014)
17. Liao, C.S., Lu, K., Baym, M., Singh, R., Berger, B.: Isorankn: spectral methods for global alignment of multiple protein networks. Bioinformatics **25**(12), 253–8 (2009)
18. Liben-Nowell, D., Kleinberg, J.: The link prediction problem for social networks. In: CIKM, p. 13451347 (2003)

19. Liu, Y., Li, Z., Xiong, H., Gao, X., Wu, J.: Understanding of internal clustering validation measures. In: ICDM, pp. 911–916 (2010)
20. Lock, E.F., Dunson, D.B.: Bayesian consensus clustering. Bioinformatics **29**(20), 2610–2616 (2013)
21. Lourenço, A., Bulò, S.R., Rebagliati, N., Fred, A.L., Figueiredo, M.A., Pelillo, M.: Probabilistic consensus clustering using evidence accumulation. Mach. Learn. **98**(1), 331–357 (2015)
22. Lu, C.T., Shuai, H.H., Yu, P.S.: Identifying your customers in social networks. In: Conference on Information and Knowledge Management, pp. 391–400 (2014)
23. Luxburg, U.: A tutorial on spectral clustering. Stat. Comput. **17**(4), 395–416 (2007)
24. Malliaros, F.D., Vazirgiannis, M.: Clustering and community detection in directed networks: a survey. Phys. Rep. **533**(4), 95–142 (2013)
25. Newman, M.E., Girvan, M.: Finding and evaluating community structure in networks. Phys. Rev. E **69**(2), 026,113–026,113 (2004)
26. Nguyen, N., Caruana, R.: Consensus clusterings. In: ICDM, pp. 607–612 (2007)
27. Qi, G.J., Aggarwal, C.C., Huang, T.: Community detection with edge content in social media networks. In: 2012 IEEE 28th International Conference on Data Engineering, pp. 534–545 (2012)
28. Shao, W., Zhang, J., He, L., Yu, P.S.: Multi-source multi-view clustering via discrepancy penalty. CoRR abs/1604.04029 (2016)
29. Shi, J., Malik, J.: Normalized cuts and image segmentation. TPAMI **22**(8), 888–905 (2000)
30. Singh, R., Xu, J., Berger, B.: Pairwise global alignment of protein interaction networks by matching neighborhood topology. In: RECOMB, pp. 16–31 (2007)
31. Sun, Y., Barber, R., Gupta, M., Aggarwal, C.C., Han, J.: Co-author relationship prediction in heterogeneous bibliographic networks. In: ASONAM, pp. 121–128 (2011)
32. Sun, Y., Han, J., Yan, X., Yu, P.S., Wu, T.: Pathsim: meta path-based top-k similarity search in heterogeneous information networks. Proc. VLDB Endow. **4**(11), 992–1003 (2011)
33. Sun, Y., Aggarwal, C.C., Han, J.: Relation strength-aware clustering of heterogeneous information networks with incomplete attributes. Comput. Sci. **5**(5), 394–405 (2012)
34. Sun, Y., Han, J., Aggarwal, C.C., Chawla, N.V.: When will it happen?: relationship prediction in heterogeneous information networks. In: WSDM, pp. 663–672 (2012)
35. Yin, X., Han, J., Yu, P.S.: Crossclus: user-guided multi-relational clustering. Data Min. Knowl. Disc. **15**(3), 321–348 (2007)
36. Yu, P.S., Zhang, J.: Mcd: mutual clustering across multiple social networks. In: IEEE BigData, pp. 762–771 (2015)
37. Yu, X., Sun, Y., Norick, B., Mao, T., Han, J.: User guided entity similarity search using meta-path selection in heterogeneous information networks. In: CIKM, pp. 2025–2029 (2012)
38. Zafarani, R., Liu, H.: Connecting users across social media sites: a behavioral-modeling approach. In: KDD, pp. 41–49 (2013)
39. Zhan, Q., Zhang, J., Wang, S., Philip, S.Y., Xie, J.: Influence maximization across partially aligned heterogenous social networks. In: PAKDD, pp. 58–69 (2015)
40. Zhan, Q., Zhang, J., Yu, P.S., Emery, S., Xie, J.: Discover tipping users for cross network influencing. In: IRI, pp. 67–76 (2016)
41. Zhang, J., Chen, J., Zhu, J., Chang, Y., Yu, P.S.: Link prediction with cardinality constraint. In: WSDM (2017)
42. Zhang, J., Kong, X., Yu, P.S.: Predicting social links for new users across aligned heterogeneous social networks. In: ICDM, pp. 1289–1294 (2013)
43. Zhang, J., Kong, X., Yu, P.S.: Transferring heterogeneous links across location-based social networks. In: ICDM, pp. 303–312 (2014)
44. Zhang, J., Philip, S.Y.: Integrated anchor and social link predictions across social networks. In: Proceedings of the 24th International Conference on Artificial Intelligence, pp. 2125–2131 (2015)
45. Zhang, J., Shao, W., Wang, S., Kong, X., Yu, P.S.: PNA: partial network alignment with generic stable matching. In: IRI, pp. 166–173 (2015)

46. Zhang, J., Yu, P.: Multiple anonymized social networks alignment. In: ICDM, pp. 599–608 (2015)
47. Zhang, J., Yu, P.: Pct: partial co-alignment of social networks. In: WWW, pp. 749–759 (2016)
48. Zhang, J., Yu, P.S.: Community detection for emerging networks. In: SDM, pp. 127–135 (2015)
49. Zhang, J., Yu, P.S., Lv, Y.: Organizational chart inference. In: KDD, pp. 1435–1444 (2015)
50. Zhang, J., Yu, P.S., Zhou, Z.H.: Meta-path based multi-network collective link prediction. In: KDD, pp. 1286–1295 (2014)

Chapter 7
Schema-Rich Heterogeneous Network Mining

Abstract Traditional heterogeneous information network usually has simple network schema, where there are a small number of types of nodes and links and meta paths are easily enumerated. However, in many real applications, some heterogeneous information networks have a huge number of types of nodes and links, and it is hard to depict their network schema. We call this kind of networks as schema-rich heterogeneous information network. For example, knowledge graph, constructed with $< object, relation, object >$ tuples, can be considered as a schema-rich heterogeneous network, where there are usually tens of thousands of types of nodes and links. In this chapter, we introduce two data mining tasks on schema-rich heterogeneous network: link prediction and entity set expansion. Through these two tasks, we illustrate the challenges and potential solutions on mining this kind of more complex and popular heterogeneous networks.

7.1 Link Prediction in Schema-Rich Heterogeneous Network

7.1.1 Overview

Link prediction is a fundamental data mining problem that attempts to estimate the likelihood of the existence of a link between two nodes, based on observed links and the attributes of nodes. Link prediction is the base of many data mining tasks, such as data clearness and recommendation. Some works have been done to predict link existence in heterogeneous information network (HIN). As a unique semantic characteristic of HIN, meta path [24], a sequence of relations connecting two nodes, is widely used for link prediction. Utilizing the meta path, these works usually employ a two-step process to solve link prediction problem in HIN. The first step is to extract meta path-based feature vectors, and the second step is to train a regression or classification model to compute the existence probability of a link [4, 21, 23, 28]. For example, Sun et al. [21] proposed PathPredict to solve the problem of co-author relationship prediction, Cao et al. [4] proposed an iterative framework to predict multiple types of links collectively in HIN, and Sun et al. [23] modeled

© Springer International Publishing AG 2017
C. Shi and P.S. Yu, *Heterogeneous Information Network Analysis
and Applications*, Data Analytics, DOI 10.1007/978-3-319-56212-4_7

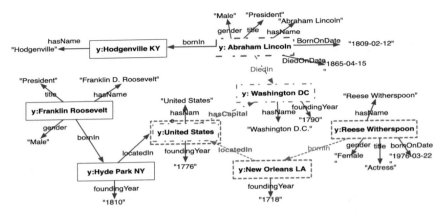

Fig. 7.1 A snapshot of the RDF structure extracted from DBpedia

the distribution of relationship building time to predict when a certain relationship will be formed. These works usually have a basic assumption: the meta paths can be predefined or enumerated in a simple HIN. When the HIN is simple, we can easily and manually enumerate some meaningful and short meta paths [24]. For example, a bibliographic network with star schema is used in [21, 23, 28] and only several meta paths are enumerated.

However, in many real networked data, the network structures are more complex, and meta paths cannot be enumerated. Knowledge graph is the base of the contemporary search engine [19], where its resource description framework (RDF) [25] $< object, relation, object >$ naturally constructs an HIN. In such an HIN, the types of nodes and relations are huge. For example, DBpedia [2], a kind of knowledge graph, has recorded more than 38 million entities and 3 billion facts. In this kind of networks, it is hard to describe them with simple schema, so we call them schema-rich HIN. Figure 7.1 shows a snapshot of the RDF structure extracted from DBpedia. You can find that there are many types of objects and links in such a small network, e.g., Person, City, and Country. Moreover, there are many meta paths to connect two object types. For example, for Person and Country types, there are two meta paths: $Person \xrightarrow{bornin} City \xrightarrow{locatedIn} Country$ and $Person \xrightarrow{Diedin} City \xrightarrow{hasCapital^{-1}} Country$. Note that Fig. 7.1 is one extreme little part of the whole DBpedia network, and there will be huge number of meta paths that can connect Person and Country in a real network. So that the meta paths in this kind of schema-rich HIN are too many to enumerate and it is hard to analyze them.

To be specific, the challenges of link prediction in schema-rich HIN are mainly from two aspects. (1) The meta path cannot be enumerated. As mentioned above, there are tens of thousands of nodes and links in such schema-rich HIN and the meta paths in the network have the same order of magnitude. It is impossible to enumerate meta paths between two node types. (2) It is also not easy to effectively integrate these meta paths. Even though masses of meta paths can be found between target nodes, most of them are meaningless or less important for link prediction. So that we need to learn weight for each meta path, where the weight represents the importance of paths for link prediction.

In this chapter, we study the link prediction in schema-rich HIN and propose the *Link* Prediction with *a*utomatic meta *P*aths method (LiPaP). The LiPaP designs a novel algorithm, called Automatic Meta Path Generation (AMPG), to automatically extract meta paths from schema-rich HIN. And then, we design a supervised method with likelihood function to learn the weights of meta paths. On a real knowledge base Yago, we do extensive experiments to validate the performances of LiPaP. Experiments show that LiPaP can effectively solve link prediction in schema-rich HIN through automatically extracting important meta paths and learning the weights of paths.

7.1.2 The LiPaP Method

In this section, we firstly define the link prediction in schema-rich HIN problem and then present a novel link prediction method named LiPaP. This method includes two steps: Firstly, we design an algorithm called AMPG to discover useful meta paths with training pairs automatically. Secondly, we use a supervised method to integrate meta paths to form a model for further prediction.

7.1.2.1 Problem Definition

Heterogeneous information network [10] is a kind of information network that includes different types of nodes and links. Traditional HIN usually has a simple network schema, such as bipartite [29] and star schema [17]. However, in some complex HINs, there are so many node types or link types that are hard to describe their network schema. We call the HIN with many types of nodes and links as **schema-rich HIN**. In simple HIN, the meta paths can be easily enumerated, but it is difficult to do the same in the schema-rich HIN. Data mining in schema-rich HIN will face new challenges. Specifically, we define a new task as follows:

Link prediction in schema-rich HIN. Given a schema-rich HIN G and a training set of entity node pairs $\phi = \{(s_i, t_i)|1 \leq i \leq k\}$, search a set of meta paths $\Upsilon = \{\prod_i |1 \leq i \leq e\}$ which can exactly describe the pairs. With these meta paths, we design a model $\eta(s, t|\Upsilon)$ to do link prediction on the test set $\psi = \{(u_i, v_i)|1 \leq i \leq r\}$.

7.1.2.2 Automatic Meta Path Generation

In order to extract the appropriate and relevant meta paths as model features for link prediction, we would like to show the AMPG algorithm, which can generate useful

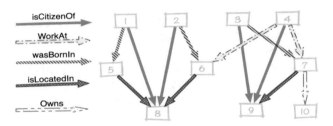

Fig. 7.2 Subgraph example of schema-rich HIN

meta paths smartly in schema-rich HIN. We would illustrate AMPG through a toy example in Fig. 7.2, where the training pairs are $\{(1, 8), (2, 8), (3, 9), (4, 9)\}$.

The main goal of AMPG is, given the training set of entity pairs, to find all the useful and relevant meta paths connecting them. These paths which to be found would not only connect more training pairs, but also show much closer relationship to present implicit features of the training set. For example, $\xrightarrow{isCitizenOf}$ is the meta path initially found by our method in Fig. 7.3, and it is not only the shortest relation but also the one connecting most training pairs. Besides, the meta paths to be found are still most relevant in the candidate paths. Basically, we start to search from the source nodes step by step to find out the useful meta paths greedily. At each step, we select the meta path that is most relevant and maybe reaching more target nodes. Then we check whether the path connects the training pairs or not. If so, we pick out the meta path, otherwise make a move forward until the unchecked meta paths are irrelevant enough. It guarantees that the generated meta paths all well describe the relationship between each training pair and the selected paths are not too many to add noise paths.

The AMPG method is a greedy algorithm that heuristically chooses the optimal paths at each step. For judging the priority of meta paths for selection, AMPG utilizes a similarity score S as a selection criterion based on a similarity measurement Path-Constrained Random Walk (PCRW) [11], which is to calculate the relevance between the given entity pairs in the meta paths. A meta path with the higher the similarity score S is more likely to be chosen.

Specifically, in AMPG, we use a data structure to record the situation of each step. The structure records a meta path passed by, a set of entity pairs reached and their PCRW values, and the similarity score S of the current structure, as shown in Fig. 7.3. Besides, we create a candidate set to record the structure to be handled.

The similarity score S of the structure mentioned above is for judging the priority of the structure. S measures the similarity of the whole arrival pairs in the structure. The highest S means the most relevant relationship and the most promising meta paths, so we get the structure with the highest S at every step. The definition of similarity score S is as follows:

$$S = \sum_s \frac{1}{T} \sum_t [\sigma(s, t | \textstyle\prod) \bullet r(s)], \qquad (7.1)$$

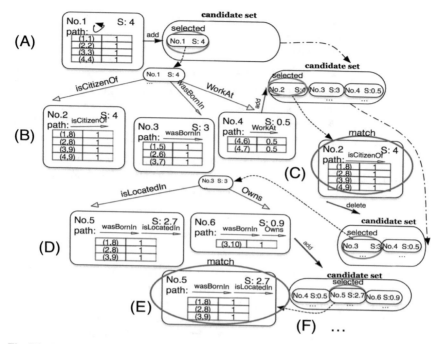

Fig. 7.3 An example of meta-path automatic generation

where s and t are source and reaching entity node, respectively, on meta path $\prod$, T is the number of reaching entity nodes, and $\sigma(s, t|\prod)$ is the PCRW value. $r(s) = 1 - \alpha \bullet N$ is the contribution of s to the current structure for training pairs selection balance, where α is the decreasing coefficient of the contribution as 0.1 because of the good performance on it, and N is the number of the target nodes that s has reached through other selected paths. It means, if one of source nodes in $\sum_s$ has more target nodes matched before, N will be larger and S will be reduced due to the smaller $r(s)$. So that the structure with other source nodes which have fewer matches will get high priority to be traversed greedily.

In order to get rid of the unimportant or the low-pair-matched meta paths, we set a threshold value l to judge the structures whether being put to the candidate set or not.

$$l = \varepsilon \bullet |A|, \tag{7.2}$$

where ε is a limited coefficient and $|A|$ is the number of entity pairs in the structure. If S is no less than l, add this structure into the candidate set, otherwise delete it.

Furthermore, we explain AMPG with a case study shown in Fig. 7.3. The training pairs are (1, 8), (2, 8), (3, 9), and (4, 9) and sources nodes are 1, 2, 3, and 4. The case starts with creating an initial structure No.1 and inserts it into the candidate set as shown in Fig. 7.3a. The entity pair is composed of the source node and itself and no meta path is generated at this step. Our algorithm will read candidate set iteratively and choose the structure with highest S at each step. For each selected structure, it will be checked if any training pairs are matched. If not, we move one step in HIN, as shown in Fig. 7.3b. We can pass by three edge types $\xrightarrow{isCitizenOf}$, $\xrightarrow{wasBornIn}$, and $\xrightarrow{WorkAt}$. For each passed edge type, we create new structures such as No.2 and No.4. Then, we check the new structures whether fit the conditions of expanding further and insert them into the candidate set. Remove the used structure No.1 and read next structure. Otherwise, as shown in Fig. 7.3c, four pairs are matched, so a new relevant meta path $\xrightarrow{isCitizenOf}$ is generated and its similarity value vector is recorded. Remove the used structure No.2 and continue to read next. The algorithm terminates when the candidate set is empty. The detail process of AMPG is found in [6].

7.1.2.3   Integration of Meta Paths

Each meta path found by AMPG is important but has different importances for further link prediction. It is necessary to find a solution of measuring the importance for each meta path and integrating them into a link prediction model.

The link prediction can be considered as a classification problem. So we use the positive and negative samples to train a model to predict whether the link exists between the given pairs or not. Positive samples are the training pairs, while negative samples are generated by replacing the target nodes of the training pairs with the same-typed nodes without the same relations. Thus, a positive value is the similarity value vector of each positive pair on all selected meta paths, while a negative value is the vector of negative pair.

For training model, we assume that the weight of each meta path $\prod_i$ is $\varpi_i(i = 1, \cdots, N)$, $\varpi_i \geq 0$, and $\sum_{i=1}^{N} \varpi_i = 1$. In order to train the appropriate path weights, we use the log-likelihood function. The specific formula is as follows:

$$\max h = \sum_{x^+ \in q^+} \frac{\ln(t(\varpi, x^+))}{|q^+|} + \sum_{x^- \in q^-} \frac{\ln(1 - t(\varpi, x^-))}{|q^-|} - \frac{||\varpi||^2}{2}, \qquad (7.3)$$

where $t(\varpi, x)$ is the Sigmoid function (i.e., $t(\varpi, x) = \dfrac{e^{\varpi^T x}}{e^{\varpi^T x} + 1}$). x is similarity value vector of sample pair in all selected paths, x^+ is positive sample, and x^- is negative sample. q^+ is similarity matrix of positive pairs made of x^+. And q^- is similarity matrix of negative pairs made of x^-. $\dfrac{||\varpi||^2}{2}$ is the regularizer to avoid overfitting.

After learning weights of relevant meta paths Υ, we use a logistic regression model to integrate meta paths for link prediction.

$$\eta(s, t | \Upsilon) = (1 + e^{-(\sum_{x \in \Upsilon} \varpi_x \bullet \sigma(s, t | \prod_x) + \varpi_0)})^{-1}, \qquad (7.4)$$

where (s, t) is the pair we should do link prediction, and x is each selected meta path feature, while ϖ_x is the weight of x we learn above. And Υ is the set of selected meta paths. If $\eta(s, t|\Upsilon)$ is larger than a specific value, we judge they would be connected by the link predicted.

7.1.3 Experiments

In order to verify the superiority of our designed method of link prediction in schema-rich HIN, we conduct a series of relevant experiments and validate the effectiveness of LiPaP from four aspects.

7.1.3.1 Experiment Settings

In our experiments, we use Yago to conduct relevant experiments and it is a large-scale knowledge graph, which is derived from Wikipedia, WordNet, and GeoNames [20]. The dataset includes more than ten million entities and 120 million facts made from these entities. We only adopt "*COREFact*" of this dataset, which contains 4484914 facts, 35 relationships, and 1369931 entities of 3455 types. A fact is a triple: $< entity, relationship, entity >$, e.g., $< NewYork, locatedin, UnitedStates >$.

We use receiver operating characteristic curve known as ROC curve to evaluate the performance of different methods. It is defined as a plot of true positive rate (TPR), as the y coordinate versus false positive rate (FPR), and as the x coordinate. TPR is the ratio of the number of true positive decisions and actual positive cases while FPR is the ratio of the number of false positive decisions and actual negative cases. The area under the curve is referred to the AUC. The larger the area is, the larger the accuracy in prediction is.

7.1.3.2 Effectiveness Experiments

This section will validate the effectiveness of our prediction method LiPaP on accurately predicting links existing in entity pairs. Since there are no existing solutions for this problem, as a baseline (called PCRW [11]), we enumerate all meta paths, and the same weight learning method with LiPaP is employed. Because meta paths with length more than 4 are most irrelevant, the PCRW enumerates the meta paths with the length no more than 1, 2, 3, and 4, and the corresponding methods are called PCRW-1, PCRW-2, PCRW-3, and PCRW-4, respectively. Based on Yago dataset, we randomly and, respectively, select 200 entity pairs from two relations $\xrightarrow{isLocatedIn}$ and $\xrightarrow{isCitizenOf}$. Note that, we assume that these two types of links are not available in the prediction task. In this experiment, 100 entity pairs of them are used as the training set; the others are used as the test set. In LiPaP, we set ε in Eq. 7.2 as 0.005 and the max path length is also limited to 4.

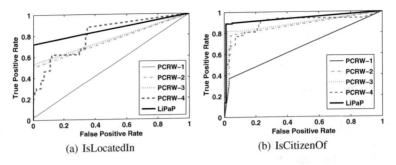

Fig. 7.4 Prediction accuracy of different methods on two link prediction tasks

Table 7.1 Most relevant 4 meta paths for *isCitizenOf*

Meta path	Weight
Person $\xrightarrow{wasBornIn}$ City $\xrightarrow{islocatedIn}$ County	0.1425
Person $\xrightarrow{livesIn}$ County	0.0819
Person $\xrightarrow{livesIn}$ City $\xrightarrow{islocatedIn}$ County	0.0744
Person $\xrightarrow{wasBornIn}$ City $\xleftarrow{isLeaderOf}$ Person $\xrightarrow{graduatedFrom}$ university $\xrightarrow{islocatedIn}$ County	0.0609

The results of two link prediction tasks are shown in Fig. 7.4. It is clear that LiPaP has better performances than all PCRW methods, which implies that LiPaP can effectively generate useful meta paths. Moreover, the PCRW generally has better performance when the path length is longer, since it can exploit more useful meta paths. However, it will take more cost to search more meta paths, most of which are irrelevant. For example, PCRW-3 generates more than 80 paths and PCRW-4 finds more than 600 paths with lots of irrelevant paths. On the contrary, LiPaP only generates 30 meta paths for the $\xrightarrow{isCitizenOf}$ task.

In order to intuitively observe the effectiveness of meta paths found, Table 7.1 shows the top four generated meta paths and the corresponding training weights for the $\xrightarrow{isCitizenOf}$ task. It is obvious that four meta paths are all relevant to the link $\xrightarrow{isCitizenOf}$. The most relevant one is the first meta path which shows the fact that a person is born in a city and the city is located in a country. It describes the citizen relationship in fact. The last one with length 4 seems not to be close, but actually has certain logistic relation with the link $\xrightarrow{isCitizenOf}$. However, these long and important meta paths can be missed if the maximum length of meta path was limited too short, as PCRW does. While our method can automatically find these paths and assign them a high importance.

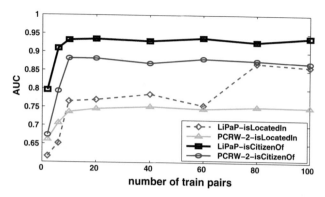

Fig. 7.5 Influence of different sizes of training set

7.1.3.3 Influence of the Size of Training Set

In this section, we evaluate the influence of the size of training set on the prediction performances. The sizes of training set are set with {2, 6, 10, 20, 40, 60, 80, 100}. Besides our LiPaP, we choose PCRW-2 as baseline, since it can generate most of useful meta paths and achieve good performances compared to other PCRW methods. As illustrated in Fig. 7.5, when the number of training pairs is smaller than 10, the performances of both methods improve rapidly with the size of pairs growing. However, when the size is more than 10, the size of training set has little effect on the performances of both methods. We think the reason lies in that too small training set cannot discover all useful meta paths, while large training set may introduce much noise. When the size of training set is from 10 to 20 in this dataset, it is good enough to discover all useful meta paths and avoid much noise. Furthermore, it can save space and time to learn model and make the performance of our method better.

7.1.3.4 Impact of Weight Learning

To illustrate the benefit of weight learning, we redone the experiments on the $\xrightarrow{isCitizenOf}$ task mentioned above. We run LiPaP with the weight learning or random weights, and with average weights. Figure 7.6 shows the performances of these methods. It is obvious that the weight learning can improve prediction performances. The model with random weight performs worst, owing to giving the more relevant paths low weights. The model with weight just has a little better performance than the model with average weight, because the meta path features generated by AMPG are all relevant and important, the most important feature also has not got a very low weight in the model with average weight. So the performance of the model with average weight is also not poor in spite of being inferior to the model with weight. Therefore, the weight learning can adjust the importance of different meta paths so as to integrate them well and make the model better.

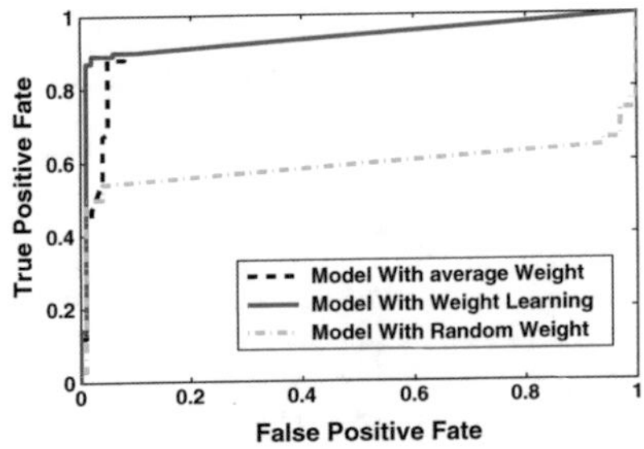

Fig. 7.6 Effectiveness of
weight learning

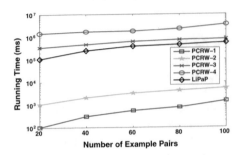

Fig. 7.7 Running times of
different methods

7.1.3.5 Efficiency Experiments

In this section, we choose five different sizes of training set, i.e., {20, 40, 60, 80, 100},
to validate the efficiency of finding meta paths of different methods. Figure 7.7
demonstrates the running time on different models for the $\xrightarrow{isLocatedIn}$ task. It is obvious that the running times of these models approximate an linear increase with the
increase in the size of the training set. In spite of the small running time, the short
meta paths found by PCRW-1 and PCRW-2 restrict their prediction performances.
Our LiPaP has smaller running time than PCRW-3 and PCRW-4, since it only finds
a small number of important meta paths. In this way, LiPaP has a better balance on
effectiveness and efficiency.

7.2   Entity Set Expansion with Meta Path in Knowledge
    Graph

7.2.1   Overview

Entity Set Expansion (ESE) refers to the problem of expanding a small set with a few
seed entities into a more complete set, entities of which belong to a particular class.
For example, given a few seeds like "China," "America," and "Russia" of country
class, ESE will leverage data sources (e.g., text or Web information) to obtain other

country instances, such as Japan and Korea. ESE has been used in many applications, e.g., dictionary construction [7], query refinement [9], and query suggestion [5].

Amounts of methods have been proposed for ESE and most of them are based on the text or Web environment [8, 12, 16, 26, 27]. These methods utilize distribution information or context pattern of seeds to expand entities. For instance, Wang and Cohen [26] propose a novel approach that can be applied to semi-structured documents written in any markup language and in any human language. Recently, knowledge graph has become a popular tool to store and retrieve fact information with graph structure, such as Wikipedia and Yago. Among those texts or Web based methods, some ones also began to leverage knowledge graph as auxiliary for the performance improvement of ESE. For example, Qi et al. [15] use Wikipedia semantic knowledge to choose better seeds for ESE. However, seldom work only utilizes knowledge graph as an individual data source for ESE.

In this chapter, we firstly study the entity set expansion with knowledge graph. Since knowledge graph is usually constituted by $< object, relation, object >$ tuples, we can consider it as a heterogeneous information network (HIN) [18], which contains different types of objects and relations. Based on this HIN, we design a *M*eta *P*ath based *E*ntity *S*et *E*xpansion approach (called MP_ESE). Specifically, the MP_ESE employs the meta path [22], a relation sequence connecting entities, to capture the implicit common feature of seed entities, and designs a Seed-based Meta Path Generation method, called SMPG, to exploit the potential relations among entities. In addition, a heuristic weight learning method is adopted to assign the importance of meta paths. With the help of weighted meta paths, MP_ESE can automatically extend entity set. Based on the Yago knowledge graph, we generate four different types of entity set expansion tasks. On almost all tasks, the proposed method outperforms other baselines.

7.2.2 The MP_ESE Method

In order to solve the problem of ESE with knowledge graph, we propose a novel approach calledMP_ESE. As we have said, KG is a natural HIN, we employ the widely used meta path in HIN to exploit the potential common feature of seeds. The MP_ESE includes the following three steps. Firstly, we design a strategy of extracting candidate entities. Secondly, we develop an algorithm, called SMPG, to automatically discover important meta paths between seeds. Finally, we get a ranking model through combining the meta paths with a heuristic strategy.

7.2.2.1 Knowledge Graph as a HIN

Knowledge graph (KG) [19] is a large and complex graph dataset, which consists of triples of the form $< Subject, Property, Object >$, such as $< StevenSpielberg, directed, WarHorse(film) >$ shown in Fig. 7.8. Yago [20], DBpedia [1] and Freebase

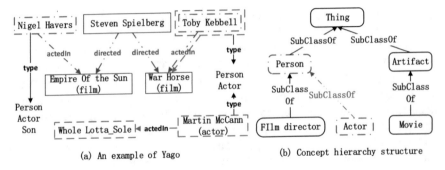

Fig. 7.8 An example of Yago with concept hierarchy structure

[3] are prime examples of KG. The types of entities or relations in KG are often organized as concept hierarchy structure, which describes the subclass relationship among entity types or relations. Figure 7.8b is a toy example and we can see that actor is subclass of person. All the types share a common root called thing.

Heterogeneous information network (HIN) [22] is a network including different types of nodes or links. In HIN, meta path [22], a sequence of relations between objects, is widely used to capture the rich semantic meaning. Since KG contains different types of objects (i.e., subject and object) and links (i.e., property), KG is a natural HIN. In Fig. 7.8, *actedIn* and *directed* are two kinds of link types and actor and film director are different object types. $Person \xrightarrow{actedIn} Movie \xrightarrow{directed^{-1}} Person$ is a meta path between Toby Kebbell and Steven Spielberg and $directed^{-1}$ is the opposite direction of the edge *directed*. In addition, Toby Kebbell and Martin McCann belong to the actor class. Toby Kebbell and Nigel Havers are not only the instances of actor class but also included in the actors who acted in movies Steven Spielberg directed. In order to distinguish the two kinds of sets, we call the latter as the fine-grained set and the former as the coarse grained set.

7.2.2.2 Candidate Entities Extraction

Because the number of entities in knowledge graph is extremely huge, it is unpractical and unreasonable to compute the similarity of each entity and seed. In order to reduce the number of candidate entities, we design a strategy, which leverages concept hierarchy structure introduced above, to get a proper set of candidate entities from knowledge graph. Specifically, it includes the following four steps as shown in Fig. 7.9. Step 1 obtains entity types of each seed. Step 2 generates the initial candidates types by the intersection operation. Step 3 filters the initial candidates types with the concept hierarchy structure. Step 4 extracts candidate entities of satisfying the ultimate candidates types.

In order to clearly illustrate the process of candidate entities extraction, we take Fig. 7.8 as an example and choose Toby Kebbell and Nigel Havers as the seeds.

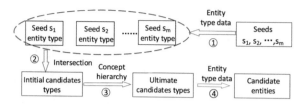

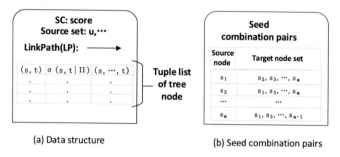

Fig. 7.9 The procedure of candidate entities extraction

Fig. 7.10 Notation of data structure and seed combination pairs

Their entity types set is {person, actor} and {son, person, actor}, respectively. And the intersection of them is {person, actor} called the initial candidates types. These candidates types may be noisy, which makes the number of candidate entities large. Therefore, we filter some candidates types using concept hierarchy structure as shown in Fig. 7.8b. We choose the class closest to the bottom as the ultimate candidates types. Here, we choose actor class. According to the ultimate types, we extract the candidate entities from Yago.

7.2.2.3 Seed-Based Meta Path Generation

In order to automatically discover meta paths between seeds, we design the Seed-based Meta Path Generation (SMPG) algorithm. The basic idea is that SMPG begins to search the KG from all seeds and finds important meta paths that connect certain number of seed pairs, and the meta paths can reveal the implicit common character of seeds.

The process of meta path generation is traversing the KG indeed, and thus a tree structure is introduced in SMPG. SMPG works by expanding the tree structure and Fig. 7.10a shows the data structure of each tree node, which stores a tuple list of entity pairs with similarity value and the set of being visited entities. The tuple form of the list is $\langle (s, t), \sigma(s,t \mid \prod), (s, \cdots, t) \rangle$, where (s, t) denotes the source node and target node of the current path $\prod$. Each tree edge denotes the link type between entities. The root node of the tree contains all entity pairs composed of each seed and itself. SMPG starts to expand from the root node step by step to discover important meta

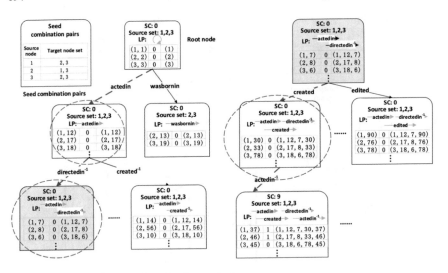

Fig. 7.11 Seed-based meta path generation method

paths. At each step, we check whether the score *SC* of the tree node is larger than the predefined threshold value *ν*, which guarantees that the meta path is important enough to reveal the character of seeds. If so, we pick out the corresponding meta path, otherwise make a move forward until the tree can not be further expanded. When moving forward, we choose the tree node with the maximum number of different source nodes as well as the minimum number of tuples to expand, which indicates that the path of the tree node covers more seeds and has a better discriminability.

Specifically, in SMPG, we use a source set in the tree node to record the source nodes of all entity pairs in the tuple list. In order to prevent the circle, we record the nodes having been visited along the path $\prod$ in $(s, \cdots , t)$ of the tuple $\langle (s, t), \sigma(s,t|\prod), (s, \cdots , t) \rangle$. Here, $\sigma(s,t|\prod)$ is the similarity that represents whether node t is in the target node set of source node s, it is 1 if so and 0 otherwise. The target node set of each source node can be found in seed combination pairs as shown in Fig. 7.10b and each seed can be combined with the other seeds. $\sigma(s,t|\prod)$ also means that whether the meta path connects the seed pair. And seed pairs that each meta path connects are also recorded. In addition, *LP* is the passing link path and the score *SC* of the tree node is the sum of all tuples similarity, which measures the importance of the tree node or path.

Let us elaborate the process with an example shown in Fig. 7.11, where the set of seeds is {Toby Kebbell, Nigel Havers, Harrison Ford} marked as {1, 2, 3}. The set of seed combination pairs is {[1, (2, 3)], [2, (1, 3)], [3, (1, 2)]} shown in Fig. 7.11. The root node of the tree contains all entity pairs composed of each seed and itself, and has *SC* = 0. The first expansion passes through two types of links: *actedIn* and *wasBornIn*, and gets two new tree nodes. For each new tree node, SMPG records each tuple, *P* and *SC* as well as source set. At the moment, all paths do not connect any seed pairs, so we choose the tree node with the maximum number of source set as well as the minimum number of tuples to expand. Here, we choose the tree node with link *actedIn* to expand and then get five new tree nodes. Figure 7.11 only demonstrates two of them. After the second expansion, there is not still path connecting seed pairs. Then we continue to choose the tree node with the maximum number of source set and the minimum number of tuples to expand, and we update the corresponding values. After several expansions, a length-4 path $Actor \xrightarrow{actedIn} Movie \xrightarrow{directed^{-1}} Person \xrightarrow{created} Movie \xrightarrow{actedIn^{-1}} Actor$ is found shown by the dash line in Fig. 7.11. And we continue to repeat the process until the condition is satisfied or the tree can not be further expanded.

7.2.2.4 Expanding Entities with Meta Paths

SMPG discovers the important meta paths P, but the importance of each meta path is different for the further entity set expansion, and it is related to the number of seed pairs that meta path connects. Intuitively, the more seed pairs the meta path connects, the more important it is. Thus, we consider the ratio of SP_k and $m * (m - 1)$ to be the weight w'_k of meta path $p_k(p_k \in P)$, where SP_k is the number of seed pairs that meta path P_k connects, $m * (m - 1)$ denotes the total number of seed pairs, and m is the number of seeds. In order to normalize w'_k, we define the final weight as follows:

$$w_k = \frac{w'_k}{\sum_{k=1}^{l} w'_k} \tag{7.5}$$

where l is the number of meta paths P.

With the w_k, we can combine meta paths to get the following ranking model:

$$R(c_i, S) = \frac{1}{m} \sum_{j=1}^{m} \sum_{k=1}^{l} w_k \cdot r\{(c_i, s_j)|p_k\} \quad s_j \in S, \ i \in \{1, 2, \cdots, n\} \tag{7.6}$$

where c_i denotes the ith candidate entity and n is the number of candidates. $S = \{s_1, s_2, \cdots, s_m\}$ is the set of seeds and l is the number of meta paths. $r\{(c_i, s_j)|p_k\}$ denotes whether the path p_k connects c_i and s_j; it is 1 if connected and 0 otherwise.

We can compute relevance between each candidate entity and each seed using the ranking model in Eq. 7.6, and then rank all candidate entities.

7.2.3 Experiments

7.2.3.1 Experiment Settings

As a typical KG, Yago [20] has knowledge about more than ten million entities and contains more than 120 million facts. We adopt "yagoFacts," "yagoSimpleTypes,"

Table 7.2 Description of the data

Data	Template of triples	# triples
yagoFacts	< entity relatinship entity >	4,484,914
yagoSimpleTypes	< entity rdf:type wordnet_type >	5,437,179
yagoTaxonomy	< wordnet_type rdfs:subclassof wordnet_type >	69,826

and "yagoTaxonomy" parts of this dataset to conduct experiments, which contain 35 relationships, more than 1.3 million entities of 3455 instance classes. Table 7.2 is the description of the relevant data.

We choose four representative expansion tasks to evaluate the performance of MP_ESE. The classes used in these tasks are summarized as follows: Actors of the movies Steven Spielberg directed, softwares of the companies located in Mountain View of California, movies whose director won National Film Award, and scientists of the universities located in Cambridge of Massachusetts. Four classes are written as Actor*, Software*, Movie*, and Scientist*, the real number of instances in these four classes are 112, 98, 653, and 202, respectively.

We employ two popular criteria of precision-at-k ($p@k$) and mean average precision (MAP) to evaluate the performance of our approach. $p@k$ is the percentage of top k results that belong to correct instances. Here, they are $p@30$, $p@60$, and $p@90$. MAP is the mean of the average precision (AP) of the $p@30$, $p@60$, and $p@90$. $AP = \frac{\sum_{i=1}^{k} p@i \times rel_i}{\text{# of correct instances}}$, where rel_i equals 1 if the result at rank i is correct instance and 0 otherwise.

7.2.3.2 Effectiveness Experiments

In this section, we will validate the effectiveness of MP_ESE on entity set expansion. Since there are no direct solutions for ESE on KG, we design the following three baselines:

- Link-Based. According to the pattern-based methods in text or Web environment, we only consider 1-hop link of an entity, denoted as Link-Based.
- Nearest-Neighbor. Inspired by QBEES [13, 14], we consider 1-hop link and 1-hop entity at the same time, called Nearest-Neighbor.
- PCRW. Based on the path-constrained random walk [11], we only compare with length-2 path, denoted as PCRW. The reason is that the longer path needs more running time.

For each class introduced above, we randomly take three seeds from the instance set to conduct an experiment. We run algorithms 30 times and record the average results. In MP_ESE, we set the predefined threshold value v to be $m * (m - 1)/2 + 1$, which can guarantee that the path connects half number of seeds or more, m is the number of seeds. And the max length of path is set to be 4 since meta paths with

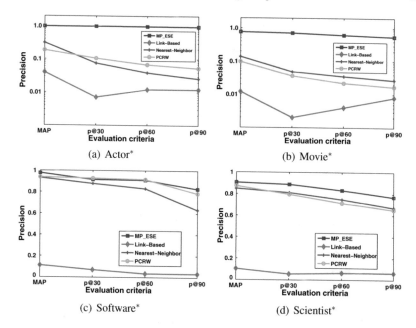

Fig. 7.12 The result of entity set expansion

length more than 4 are almost irrelevant. The optimal parameters are set for other baselines.

The overall results of entity set expansion are given in Fig. 7.12. From Fig. 7.12, we can see that our MP_ESE approach achieves better performances than other methods on almost all conditions, especially on the Actor* and Movie* tasks. All baselines have very bad performances on Actor* and Movie*. We think the reason is that the 1-hop link or 1-hop entity cannot further distinguish the character of the fine-grained class but MP_ESE can distinguish them well. On the Software* task, MP_ESE and PCRW have close performance. The reason is that Software* is an overlapping class and has another class label depicted by length-2 path $Software \xrightarrow{created^{-1}} Company \xrightarrow{created} Software$. Due to the fact that it has few semantic meaning, Link-Based has very bad performance. In all, MP_ESE has the best performances because it employs the important meta paths between seeds and can capture the subtle semantic meaning.

In order to intuitively observe the effectiveness of discovered meta paths, Table 7.3 depicts the top three meta paths returned by SMPG for Actor*. We observe that these meta paths reveal some common traits of Actor*. The first meta path indicates that actors act in movies directed by the same director, which shows that SMPG can effectively mine the most important semantic meaning of Actor*. The second and the third meta paths imply that some actors act in movies edited or composed by the same person. Through leveraging the important meta paths discovered by SMPG, we can find other entities belonging to the same class with seeds.

Table 7.3 Most relevant 3 meta paths for Actor*

Meta path	w
Person $\xrightarrow{actedIn}$ Movie $\xrightarrow{directed^{-1}}$ Person $\xrightarrow{directed}$ Movie $\xrightarrow{actedIn^{-1}}$ Person	0.2180
Person $\xrightarrow{actedIn}$ Movie $\xrightarrow{writeMusicFor^{-1}}$ Person $\xrightarrow{writeMusicFor}$ Movie $\xrightarrow{actedIn^{-1}}$ Person	0.1495
Person $\xrightarrow{actedIn}$ Movie $\xrightarrow{edited^{-1}}$ Person $\xrightarrow{edited}$ Movie $\xrightarrow{actedIn^{-1}}$ Person	0.1476

7.3 Conclusions

In this chapter, we extend the traditional heterogenous network to the schema-rich heterogeneous network where there are a huge number of types of nodes and links, such as knowledge graph. In this kind of networks, it is difficult to depict the network schema and impossible to enumerate the potential meta paths. We study two data mining tasks in schema-rich heterogeneous networks. In the link prediction task, we design the LiPaP to predict potential links among nodes, and we also propose the MP_ESE to automatically extend entity set with knowledge graph. In these methods, it is critical to efficiently and effectively discover meta paths and learning their weights. Since the knowledge graph is widely used in text analysis and search engine, when we consider the knowledge graph as heterogenous network, it will tremendously extend the study of heterogeneous network. Simultaneously, it also provides a new way for knowledge graph mining.

References

1. Auer, S., Bizer, C., Kobilarov, G., Lehmann, J., Cyganiak, R., Ives, Z.: DBpedia: A Nucleus for a Web of Open Data. Springer, Berlin (2007)
2. Bizer, C., Lehmann, J., Kobilarov, G., Auer, S., Becker, C., Cyganiak, R., Hellmann, S.: Dbpedia-a crystallization point for the web of data. Web Semant.: Sci. Serv. Agents World Wide Web **7**(3), 154–165 (2009)
3. Bollacker, K., Evans, C., Paritosh, P., Sturge, T., Taylor, J.: Freebase: a collaboratively created graph database for structuring human knowledge. In: SIGMOD, pp. 1247–1250 (2008)
4. Cao, B., Kong, X., Yu, P.S.: Collective prediction of multiple types of links in heterogeneous information networks. In: ICDM, pp. 50–59 (2014)
5. Cao, H., Jiang, D., Pei, J., He, Q., Liao, Z., Chen, E., Li, H.: Context-aware query suggestion by mining click-through and session data. In: KDD, pp. 875–883 (2008)
6. Cao, X., Zheng, Y., Shi, C., Li, J., Wu, B.: Link prediction in schema-rich heterogeneous information network. In: PAKDD, pp. 449–460 (2016)
7. Cohen, W.W., Sarawagi, S.: Exploiting dictionaries in named entity extraction: combining semi-markov extraction processes and data integration methods. In: KDD, pp. 89–98 (2004)
8. He, Y., Xin, D.: Seisa: set expansion by iterative similarity aggregation. In: WWW, pp. 427–436 (2011)
9. Hu, J., Wang, G., Lochovsky, F., Sun, J.t., Chen, Z.: Understanding user's query intent with wikipedia. In: WWW, pp. 471–480 (2009)

10. Jaiwei, H.: Mining heterogeneous information networks: the next frontier. In: SIGKDD, pp. 2–3 (2012)
11. Lao, N., Cohen, W.W.: Relational retrieval using a combination of path-constrained random walks. Mach. Learn. **81**(1), 53–67 (2010)
12. Li, X.L., Zhang, L., Liu, B., Ng, S.K.: Distributional similarity vs. pu learning for entity set expansion. In: ACL, pp. 359–364 (2010)
13. Metzger, S., Schenkel, R., Sydow, M.: Qbees: query by entity examples. In: CIKM, pp. 1829–1832 (2013)
14. Metzger, S., Schenkel, R., Sydow, M.: Aspect-based similar entity search in semantic knowledge graphs with diversity-awareness and relaxation. In: IJCWI, pp. 60–69 (2014)
15. Qi, Z., Liu, K., Zhao, J.: Choosing better seeds for entity set expansion by leveraging wikipedia semantic knowledge. In: CCPR, pp. 655–662 (2012)
16. Sarmento, L., Jijkuon, V., de Rijke, M., Oliveira, E.: More like these: growing entity classes from seeds. In: CIKM, pp. 959–962 (2007)
17. Shi, C., Kong, X., Yu, P.S., Xie, S., Wu, B.: Relevance search in heterogeneous networks. In: EDBT, pp. 180–191 (2012)
18. Shi, C., Li, Y., Zhang, J., Sun, Y., Yu, P.S.: A survey of heterogeneous information network analysis. Comput. Sci. **134**(12), 87–99 (2015)
19. Singhal, A.: Introducing the knowledge graph: things, not strings. Official google blog (2012)
20. Suchanek, F.M., Kasneci, G., Weikum, G.: Yago: a core of semantic knowledge. In: WWW, pp. 697–706 (2007)
21. Sun, Y., Barber, R., Gupta, M., Aggarwal, C.C., Han, J.: Co-author relationship prediction in heterogeneous bibliographic networks. In: ASONAM, pp. 121–128 (2011)
22. Sun, Y., Han, J., Yan, X., Yu, P.S., Wu, T.: Pathsim: Meta path-based top-k similarity search in heterogeneous information networks. VLDB **4**(11), 992–1003 (2011)
23. Sun, Y., Han, J., Aggarwal, C.C., Chawla, N.V.: When will it happen?: relationship prediction in heterogeneous information networks. In: WSDM, pp. 663–672 (2012)
24. Sun, Y., Norick, B., Han, J., Yan, X., Yu, P.S., Yu, X.: Integrating meta-path selection with user-guided object clustering in heterogeneous information networks. In: KDD, pp. 1348–1356 (2012)
25. W3C: Rdf current status. http://www.w3.org/standards/techs/rdf#w3c_all
26. Wang, R.C., Cohen, W.W.: Language-independent set expansion of named entities using the web. In: ICDM, pp. 342–350 (2007)
27. Wang, R.C., Cohen, W.W.: Iterative set expansion of named entities using the web. In: ICDM, pp. 1091–1096 (2008)
28. Yu, X., Gu, Q., Zhou, M., Han, J.: Citation prediction in heterogeneous bibliographic networks. In: SDM, pp. 1119–1130 (2012)
29. Zha, H., He, X., Ding, C.H.Q., Gu, M., Simon, H.D.: Bipartite graph partitioning and data clustering. CoRR cs.IR/0108018 (2001)

Chapter 8
Prototype System Based on Heterogeneous Network

Abstract Because of significant advantages of heterogeneous information network, it is widely used to model networked data, and many data mining tasks have been exploited on it. Besides that, many prototype systems, even real systems, have been built based on heterogeneous networks. In these systems, heterogeneous networks are constructed, stored, and operated based on real networked data, and many novel applications are designed based on heterogeneous networks. In this chapter, we introduce two prototype systems for recommendation and further give a brief review on other systems based on heterogeneous networks.

8.1 Semantic Recommender System

8.1.1 Overview

Many recommendation methods have been proposed, which can be roughly classified into two categories: content-based filtering (CB) and collaborative filtering (CF). CB analyzes correlations between the content of the items and the user's preferences [1]. CF analyzes the similarity between users or items [2]. These methods have been applied to recommender systems and achieved great success. However, these recommender systems may have the following disadvantages.

- Conventional recommender systems usually recommend similar products to users without exploring the semantics of different similarity measures. However, the similar products are often different based on similarity semantics. For example, in the movie recommendation, the similar movies based on the same actors are different from those based on the same directors. Conventional systems usually give a recommendation without considering the subtle implications of similarity semantics. The proposed system is more appealing to provide a semantic recommendation function, which will give more accurate recommendation when users know their intents.
- Conventional systems only recommend same-typed objects. However, a system may be more useful if it simultaneously recommends more related objects under different semantics. For example, when users select movies, the system not only

© Springer International Publishing AG 2017

C. Shi and P.S. Yu, *Heterogeneous Information Network Analysis and Applications*, Data Analytics, DOI 10.1007/978-3-319-56212-4_8

Fig. 8.1 An example of
heterogeneous information
network and its schema

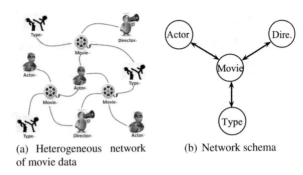

(a) Heterogeneous network   (b) Network schema
of movie data

recommends the similar movies, but also suggests some related actors and directors
(note that they are not limited to the actors and directors of this movie). The user
may find an interesting actor and then search the movies of the actor. The relevance
recommendation will provide richer information and enhance user experience.

Nowadays, social networks consisting of different types of information become
popular. Particularly, the advent of the Heterogeneous Information Network (HIN)
[4] provides a new perspective to design the recommended system. HINs are the
logical networks involving multiple-typed objects and multiple-typed links denoting
different relations. It is clear that HINs are ubiquitous and form a critical component of modern information infrastructure [4]. Although the bipartite network [8] has
been applied to organize components of recommended system, HIN is a more general
model which contains more comprehensive relations among objects and much richer
semantic information. Figure 8.1a shows an HIN example on the movie recommendation data. The network includes the richer objects (e.g., movie, actor, director) and
their relations. The network structure can be represented with the star schema as
shown in Fig. 8.1b. HIN has an unique property [10, 14]: the different paths connecting two objects have different meanings. For example, in Fig. 8.1b, movies can be
connected via "Movie–Actor–Movie" (*MAM*) path, "Movie–Type–Movie" (*MTM*)
path, and so on. It is clear that the semantics underneath these paths is different. The
MAM path means that movies have the same actors, while the *MTM* path means that
movies of the same type. Here, the meta path connecting two-typed objects is defined
as relevance path [10]. Obviously, the distinct semantics under different relevance
paths will lead to different relatedness and recommendation.

Focusing on non-personalized recommendation, this chapter demonstrates a
semantic recommended system, called HeteRecom. Different from conventional recommended systems, it is based on HIN. Generally, HeteRecom has the following
unique features. (1) Semantic recommendation: The system can recommend objects
of the designated type based on the relevance path specified by users. (2) Relevance
recommendation: Besides the same-typed objects recommendation, the system can
recommend other related objects.

The implementation of HeteRecom faces the following challenges. (1) Relevance
measure of heterogeneous objects: In order to recommend the different type objects,

Fig. 8.2 The architecture of *HeteRecom* system

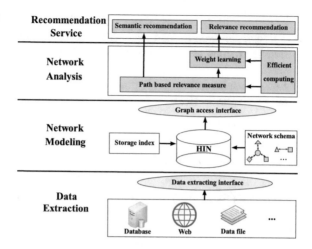

the system needs to measure the relatedness of different type objects. (2) The weight learning method: It is a key issue for an integrated recommendation to automatically determine the weights of different relevance paths. (3) Efficient computing strategies: In order to provide online service, the recommended system needs to efficiently compute the relevance measure. In order to solve these challenges, the HeteRecom system first applies a path-based relevance measure, which can not only effectively measure the relatedness of any-typed objects but also subtly capture the semantics containing in the relevance path. Besides, a heuristic weight learning method can automatically determine the weights of different paths. Moreover, many computing strategies are designed to handle huge graph data. This paper demonstrates the effectiveness of HeteRecom on the real movie data through providing online semantic and relevance recommendation services.

8.1.2 System Architecture

Figure 8.2 shows the architecture of HeteRecom, which mainly consists of four components:

- Data extraction: It extracts data from different data source (e.g., database and Web) to construct the network.
- Network modeling: It constructs the HIN with a given network schema. According to the structure of data, users can specify the network schema (e.g., bipartite, star, or arbitrary schema) to construct the HIN database. The database provides the store and index functions of the node table and edge table of the HIN.
- Network analysis: It analyzes the HIN and provides the recommendation services. It first computes and stores the relevance matrix of object pairs by the path-based relevance measure. Based on the relevance matrix and efficient computing

strategies, the system can provide the online semantic recommendation service. Through the weight learning method, it can combine the relevance information from different semantic paths and provide online relevance recommendation service.

- Recommendation service: It provides the succinct and friendly interface of recommendation services.

8.1.3 System Implementation

It is challenging in many ways to implement these components. First, it is difficult to measure the relatedness of any-typed objects in a HIN. Second, It is not easy to combine those recommendation information on different semantic paths. Third, there are many challenges in the computation and storage of huge relevance matrix. In the following section, we will present the solutions to these challenges.

8.1.3.1 A Path-Based Relevance Measure

We apply the HeteSim [10], a path-based relevance measure, to do semantic recommendation. The basic idea behind HeteSim is that similar objects are related to similar objects. The HeteSim is defined as follows:

Definition 8.1 (*HeteSim* [10]) Given a relevance path $P = R_1 \circ R_2 \circ \cdots \circ R_l$, HeteSim between two objects s and t ($s \in R_1.S$ and $t \in R_l.T$) is:

$$HeteSim(s, t|R_1 \circ R_2 \circ \cdots \circ R_l) = \frac{1}{|O(s|R_1)||I(t|R_l)|} \sum_{i=1}^{|O(s|R_1)|} \sum_{j=1}^{|I(t|R_l)|} HeteSim(O_i(s|R_1), I_j(t|R_l)|R_2 \circ \cdots \circ R_{l-1}) \tag{8.1}$$

where $O(s|R_1)$ is the out-neighbors of s based on relation R_1, $I(t|R_l)$ is the in-neighbors of t based on relation R_l, and $R.S$ ($R.T$) represents the source (target) object of relation R, respectively.

Essentially, $HeteSim(s, t|P)$ is a pairwise random walk-based measure, which evaluates how likely s and t will meet at the same node when s follows along the path and t goes against the path. The path implies the semantic information and *HeteSim* evaluates the relatedness of any-typed object pairs according to the given path. The *HeteSim* measure has shown its potential in object profiling, experts finding, and relevance search. The detailed information can be seen in [10].

Since relevance paths embody different semantics, users can specify the path according to their intents. The semantic recommendation calculates the relevance matrix with *HeteSim* and recommends the top k objects.

8.1.3.2 Weight Learning Method

There are many relevance paths connecting the query object and related objects, so the relevance recommendation should comprehensively consider the relevance measures based on all relevance paths. It can be depicted as follows:

$$Sim(A, B) = \sum_{i=1}^{N} w_i * HeteSim(A, B|P_i) \qquad (8.2)$$

where N is the number of relevance paths, P_i is a relevance path connecting the object types A and B, w_i is the weight of path P_i. Although there can be infinite relevance paths connecting two objects, we only need to consider those short paths, since the long paths are usually less important [14].

The next question is how to determine the weight w_i. The supervised learning [7] can be used to estimate these parameters. However, it is impractical for an online system: (1) It is time-consuming, even impractical, to learn these parameters on an online system. (2) It is a very labor-intensive and subjective work to label those learning instances. Here, we propose a heuristic weight learning method.

The importance (I) of a path $P = R_1 \circ R_2 \circ \cdots \circ R_l$ is determined by its strength (S) and length (l). Obviously, the path strength is decided by the strength of relations constructing the path, which can be defined as follows:

$$S(P) = \prod_{i=1}^{l} S(R_i) \qquad (8.3)$$

The strength of a relation $A \xrightarrow{R} B$ is related to the degree of A and B based on R. Intuitively, if the mutual connective links between A and B are smaller, they are more important to each other, so their relation strength is stronger. For example, the relation strength between movie and director (MD) is stronger than that between movie and type (MT). So we can define the relation strength as follows:

$$S(R) = (O(A|R)I(B|R))^{-\alpha}(\alpha \in [0, 1]) \qquad (8.4)$$

where $O(A|R)$ is the average out-degree of type A and $I(B|R)$ is the average in-degree of type B based on relation R.

The importance (I) of the path P is positively correlative to the path strength (S) and negatively correlative to the path length (l). Here, we define it as follows:

$$I(P) = f(S, l) = e^{S-l} \qquad (8.5)$$

For multiple paths $(P_1, P_2, \cdots, P_N)$, the weight w_i of path P_i is

$$w_i = \frac{I_i}{\sum_{i=1}^{N} I_i} \tag{8.6}$$

In HeteRecom, we consider all relevance paths whose length is smaller than a threshold *Len*. The relevance recommendation combines the relevance measure results of all these paths with the weight learning method and makes an integrated recommendation.

8.1.3.3 Efficient Computing Strategies

As an online recommended system, *HeteRecom* needs to do a real-time recommendation for user's query. However, an HIN is usually huge and the computation of HeteSim is time-consuming. So the system employed many efficient computing strategies. Three basic strategies are depicted as follows:

Off-line computation: The primary strategy is to compute relevance matrix offline and make recommendations online. For frequently used relevance paths, the relevance matrix $HeteSim(A, B|P)$ can be calculated ahead of time. The online recommendation on $HeteSim(a, B|P)$ will be very fast, since it only needs to locate the position in the matrix.

Fast matrix multiplications: The most time-consuming component in the system is the matrix multiplications in HeteSim. There are many frequent patterns in relevance paths. Since the matrix multiplications satisfy the associative law, we can precede to compute the product of frequent patterns iteratively. Moreover, those frequent patterns only need to be computed once. For example, we only need to compute the frequent pattern AMA once for the symmetric path $AMAMA$. Since the short pattern is more frequent, we only find the most frequent relation pair in each iteration.

Matrix sparsification: The relevance matrix often becomes denser along the matrix multiplications [7]. The dense matrix may cause two difficulties. (1) Matrix multiplications cost a lot of time and space. (2) It costs a lot of time and huge memory to load and search these dense relevance matrix. As a consequence, we need to sparsify the reachable probability matrix along the matrix multiplications without much loss of accuracy. The basic idea is to truncate those less important nodes whose relevance value is smaller than a threshold ε. The static threshold [7] is not suitable, since it may truncate some important nodes with small relevance values and keep those unimportant nodes with large relevance values. Since we usually pay close attention to the top k recommendation, we set the threshold ε as the top k relevance value of the matrix. The k is dynamically adjusted as follows:

$$k = \begin{cases} L & \text{if } L \leq W \\ \lfloor (L - W)^{\beta} \rfloor + W (\beta \in [0, 1]) & \text{others} \end{cases}$$

where L is the vector length. W is the threshold which determines the size of nonzero elements. The larger W or β may lead to the denser matrix with less loss. In order to

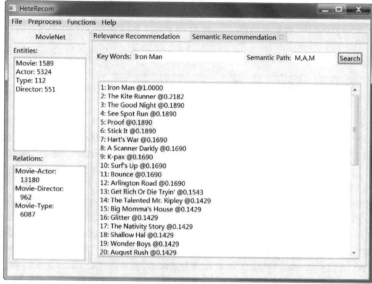

(a) Semantic recommendation based on *MAM* path

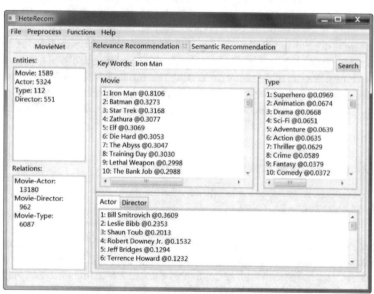

(b) Relevance recommendation

Fig. 8.3 The HeteRecom prototype system

quickly determine the top k relevance value, it is approximately computed with the sample data from the raw matrix.

8.1.4   System Demonstration

We showcase the HeteRecom prototype system using IMDB movie data as the example application. The IMDB movie data was downloaded from The Internet Movie Database.[1] The IMDB movie data collects 1591 movies before 2010. The related objects include actors, directors, and types, which are organized as a star schema shown in Fig. 8.1b.

Figure 8.3 demonstrates the interface of the HeteRecom system, which is developed with Java. The left part of interface shows the basic information of the dataset. The right part shows the recommendation results. In the semantic recommendation, users specify the key words and semantic path, the recommendation results will be exhibited in the panel. Figure 8.3a shows the movies with the same actors of "Iron Man" by specified the "MAM" path. The *HeteRecom* can make many recommendations that conventional systems cannot do. For example, recommending the movies that have the same style with the movies of "Arnold Schwarzenegger" can be done by the path *AMTM*. In the relevance recommendation, the system can simultaneously recommend different-typed objects. Figure 8.3b shows the recommendation results of the movie "Iron Man," which include the similar movies and related actors, directors, and types. We can make many interesting recommendations on *HeteRecom*. For example, if we want to know the information about the action movie, we can search "action." The system will recommend related action movies, actors, and directors.

8.2   Explainable Recommender System

8.2.1   Overview

In order to tackle the information overload problem on WWW, many recommendation techniques have been proposed to build recommender systems. These recommended systems have been widely applied to e-commerce companies and achieved great success, for example, the book recommendation in Amazon and movie recommendation in Netflix. However, the explanation of recommendation results is a very important but seldom exploited problem. Good explanations could help inspire user trust and loyalty and increase satisfaction. Recommendation explanation makes it quicker and easier for users to find what they want and persuade them to try or purchase a recommended item [20]. Contemporary explanations of recommendations

---

[1] www.imdb.com/.

Fig. 8.4 Network schema of
HIN constituted by Douban
movie recommendation

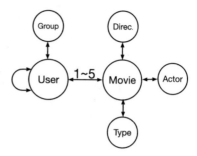

usually use features or characteristics of users or the recommended item as intermediary entities. For example, the MoviExplain system employs movie features to justify recommendations [15], and Vig et al. [21] design Tagsplanations to provide explanation based on community tags. With the surge of social recommendation, there are some works on social explanation. Wang et al. [22] propose an algorithm to generate the most persuasive social explanation; Sharma et al. [9] present a study of the effects of social explanations in a music recommendation context. These methods try to explain recommendation through one type of information (e.g., features or social relations), while the recommendation results may stem from complex heterogeneous information and various factors. The recommended system needs to explain these factors more clearly.

In this chapter, we develop a **Rec**ommender system with **Exp**lanation (called RecExp). Inspired by the recent surge of heterogeneous information network [11], we organize the objects and relations in a recommended system as an HIN. Figure 8.4 shows such an example in movie recommendation. The HIN not only contains different types of objects in movie recommendation (e.g., users and movies) but also illustrates all kinds of relations among objects, such as viewing information, social relations, and attribute information. Moreover, two objects in an HIN can be connected via different paths, called meta path, and different meta paths have different meanings. So we can find the similar users of a user through different meta paths connecting these two users, and then we can combine the recommendation results of different similar users under different meta paths. Based on this idea, we design the semantic recommended system, RecExp, with explanation, which has the following two significant features:

- Semantic recommendation: Utilizing different meta paths, RecExp can find different similar users, and thus generate different recommendation results according to these similar users. Moreover, these meta paths correspond to different recommendation models, so RecExp can realize semantic recommendation through selecting proper meta paths.
- Recommendation explanation: RecExp utilizes semantics and weights of meta paths to present personalized recommendation explanation, which can reveal user preferences and make explanation more persuasive.

8.2.2   Heterogeneous Network-Based Recommendation

In this section, we will briefly introduce the basic concept and method used in RecExp. HIN [11] is a special type of information network with the underneath data structure as a directed graph, which contains either multiple types of objects or multiple types of links. Objects and their relations in recommended system constitute an HIN. Figure 8.4 shows the network schema of the movie-recommended system in Douban, a well-known social media network in China. This movie network includes objects from six types of entities (e.g., users, movies, groups, actors) and relations between them. Links between objects represent different semantics. For example, links exist between users and users denoting the friendship relations, between users and movies denoting rating and rated relations.

8.2.2.1   Recommendation on Heterogeneous Network

For a target user, recommended systems usually recommend items according to users similar to his/her. In HIN, there are a number of meta paths [13] connecting users, such as "User–User" (UU) and "User–Movie–User" (UMU). Based on these paths, users have different types of similarities. After obtaining the path-based similarity of users, we can recommend items according to the similar users of the target user. More importantly, the meta paths connecting users have different semantics, which can represent different recommendation models. As an example shown in Fig. 8.4, the UMU path means users who view the same movies with the target user. It will recommend movies viewed by users having similar viewing records with the target user. It is collaborative recommendation in essential. Based on the HIN framework, we can flexibly represent different recommendation models through properly setting meta paths. In the following section, we will specifically introduce the semantic recommendation method, where technique details can be found in [12].

8.2.2.2   Semantic Recommendation with Single Path

Based on the path-based similarity of users, we find the similar users of a target user under a given path, and then the rating score of the target user on an item can be inferred according to the rating scores of his similar users on the item. Assume that the range of rating scores is form 1 to N (e.g., 5); P is a set of meta paths; $R \in \mathbf{R}^{|U| \times |I|}$ is the rating matrix, where $R_{u,i}$ denotes the rating score of user u on item i; and $S \in \mathbf{R}^{|U| \times |U|}$ is the path-based similarity matrix of users, where $S_{u,v}^{(l)}$ is the similarity of users u and v under path P_l. Note that the similarity matrix can be calculated offline with some path-based similarity measures [13]. Under a meta path P_l, the predicted rating score of a user u on an item i denoted as $\hat{R}_{u,i}^{(l)}$ is:

$$\hat{R}^{(l)}_{u,i} = \frac{\sum_{v=1}^{|U|} S^{(l)}_{u,v} \times R_{v,i}}{\sum_{v=1}^{|U|} S^{(l)}_{u,v}}.$$

(8.7)

According to Eq. 8.7, we can predict the rating score of a user on an item under a given path, and then recommend the item with the high score for a target user.

8.2.2.3 Hybrid Recommendation with Multiple Paths

Under different meta paths, there are different predicted rating scores. In order to calculate the composite score, we employ a personalized weight learning method with weight regularization [12]. As we know, many users have the similar interest preferences, that is, we assume that two similar users have consistent weight preferences on meta paths. For users with little rating information, their path weights can be learnt from the weights of their similar users, since the similarity information of users are more available through meta paths. So we design a weight regularization term, which compels the weights of a user to be consistent to the average of weights of his similar users. The weight matrix is denoted as $W \in R^{|U| \times |P|}$, in which each entry, denoted as $W^{(l)}_u$, means the preference weight of user u on path P_l. The column vector $W^l \in R^{|U| \times 1}$ means the weight vector of all users on path P_l. The following optimization function can learn users' preference weight W.

$$\min_W L(W) = \frac{1}{2} ||I \odot (R - \sum_{l=1}^{|P|} diag(W^{(l)}) \hat{R}^{(l)})||_2^2$$

$$+ \frac{\lambda_1}{2} \sum_{l=1}^{|P|} ||W^{(l)} - \bar{S}^{(l)} W^{(l)}||_2^2 + \frac{\lambda_0}{2} ||W||_2^2$$

(8.8)

s.t. $\qquad W \geq 0.$

where $\bar{S}^{(l)}_{u,v} = \frac{S^{(l)}_{u,v}}{\sum_v S^{(l)}_{u,v}}$ is the normalized user similarity based on path P_l, I is an indicator matrix with $I_{u,i} = 1$ if user u rated item i, and otherwise $I_{u,i} = 0$, the notation $\odot$ is the Hadamard product between matrices, and $diag(W^{(l)})$ means the diagonal matrix transformed from a vector $W^{(l)}$.

And thus, the predicted rating $\hat{R}_{u,i}$ of user u rating item i under all paths is as follows:

$$\hat{R}_{u,i} = \sum_{l=1}^{|P|} W^{(l)}_u \times \hat{R}^{(l)}_{u,i}.$$

(8.9)

The hybrid recommendation results combine the recommendation from multiple meta paths, and the weight matrix W records the user preferences on these paths. So we can explain the recommendation results according to user path preferences and semantics containing in each path.

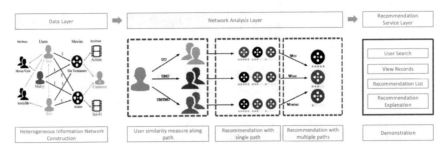

Fig. 8.5 The architecture of *RecExp* system

8.2.3   System Framework

According to the HIN-based recommendation method introduced above, we design the *RecExp* system. Figure 8.5 shows the system architecture. The three main components are detailed as follows:

- Data layer: It extracts data from different data sources (e.g., database and Web) to construct an HIN. Figure 8.4 shows the network schema of HIN in our movie-recommended system demo.
- Network analysis layer: It analyzes the HIN and provides the recommendation services. As we have illustrated in the above section, it first computes the similarities between users along different meta paths, such as "User–Movie–User." And then, based on similarity of users, we find the similar users of a target user under a given path, and the predicted rating score of the target user on a movie can be inferred from the rating scores of these similar users on the movie. Under different meta paths, there are different predicted rating scores. Through the weight learning method, we assign each meta path with a preference weight for each user, and the final predicted rating under all meta paths can be the weighted average of predicted rating under each meta path.
- Recommendation service layer: It provides the succinct and friendly Web interface of recommendation services. The recommendation services include five kinds of semantic recommendations through setting different meta paths, hybrid recommendation with explanation, and the view record for the searched user.

8.2.4   System Demonstration

Figure 8.6 demonstrates the interface of the *RecExp* system. It consists of five major components:

- **Search box**: Users can input a certain user ID in the search box.

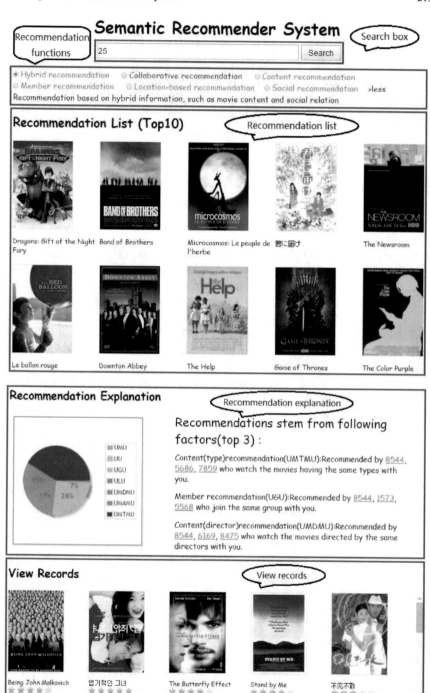

Fig. 8.6 The *RecExp* system

Table 8.1 The recommendation results of different recommendation functions

Recommendation model	Top five recommendation results
Collaborative recommendation	Sherlock, 127 Hours, Game of Thrones, Taxi Driver, The Crucible
Content recommendation	The Big Bang Theory 2, Once a Thief, The Big Bang Theory 4, 2 Broke Girls, The Monkey King
Member recommendation	The cove, Detachment, Inglorious Basterds, The Lives of Others, All About Lily
Location-based recommendation	Farewell My Concubine, Nuovo Cinema Paradise, The Cove, Saving Private Ryan, Sherlock
Social recommendation	Spirited Away, The Pursuit of Happiness, Edward, Scissor Hands

- **Recommendation functions**. There are six recommendation function buttons. Each function button represents a typical recommendation model through selecting a meta path. For example, the collaborative filtering corresponds to the UMU path, and the social recommendation corresponds to the UU path. The description of the selected recommendation model is detailed under the button box. For example, if you press the "Hybrid recommendation" button, the below panel will show "Recommendation based on hybrid information, such as movie content and social relation."

- **Recommendation list**. It shows the top 10 results recommended by the recommendation method you select.

- **Recommendation explanation**: The function will be invoked when the "Hybrid recommendation" function is selected. Since the hybrid recommendation generates the results through multiple meta paths, the fan chart shows the weights of each meta path which can represent the user preference on these paths. The larger the weight is, the more the user prefers to get recommendation from the corresponding meta path. On the right of the fan chart, it shows three most important meta paths and corresponding explanations. In each explanation, we display the three most similar users with the target user based on corresponding meta path.

- **View records**: It displays the view records of a certain user.

We showcase the *RecExp* prototype system using Douban movie data as the example application. The Douban movie data was downloaded from Douban Web site.[2] The dataset includes 13,367 users and 12,677 movies with 1,068,278 movie ratings ranging from 1 to 5, which are organized as a star-schema HIN shown in Fig. 8.4. The dataset includes the social relations among users and the attribute information of users and movies. With this dataset, we will illustrate two major functions: semantic recommendation and hybrid recommendation with explanation.

In the semantic recommendation, users can specify a user ID and the recommendation model such as collaborative recommendation, the recommendation results

---

[2] www.douban.com/.

will be exhibited in the below panel. For example, we specify user 25 and select five different recommendation functions, whose recommendation results are shown in Table 8.1. We can see that the recommendation results are different based on different meta paths. Different users have their personalized preferences. Through setting proper recommendation model, users can find their own favorite movies. For example, if a user prefers to get new movies by friends' recommendations, he can choose the social recommendation.

When we select the "Hybrid recommendation" function, the system will recommend a composite results stemming from five semantic recommendation models and display top 10 recommendation. Moreover, the recommendation explanation box will explain the recommendation reasons.

For example, we search user 25 and select the "Hybrid recommendation" function, the system will show the recommendation results and give the recommendation explanation shown in Fig. 8.6. In this case, the UMTMU path has the largest weight which means this user has the preference on a certain film type. Among movies that this user has seen, the drama and love movies are his favorite. So these types of movies make the largest proportion in his recommendation list. The system captures this user's preference for film type and displays it in the fan chart. In addition, the system displays three explanations corresponding to the three most important meta paths. For example, the system will list three most similar users with the same file type taste under the UMTMU path, if they are willing to be shown under privacy agreement.

8.3 Other Prototype Systems on Heterogeneous Network

In the section above, we have introduced two prototype systems for recommendation based on HIN. Besides that, many demo systems have also designed prototype applications on HIN. Yu et al. [23] demonstrate a prototype system on query-driven discovery of semantically similar substructures in heterogeneous networks. Danilevsky et al. [3] present the AMETHYST system for exploring and analyzing a topical hierarchy constructed from an HIN. In LikeMiner system, Jin et al. [6] introduce a heterogeneous network model for social media with "likes," and propose "like" mining algorithms to estimate representativeness and influence of objects. Meanwhile, they design SocialSpamGuard [5], a scalable and online social media spam detection system for social network security. Taking DBLP as an example, Tao et al. [18] construct a Research-Insight system to demonstrate the power of database-oriented information network analysis, including ranking, clustering, classification, recommendation, and prediction. Furthermore, they construct a semi-structured news information network NewsNet and develop a NewsNetExplorer system [19] to provide a set of news information network exploration and mining functions.

Some real application systems have also been designed. One of the most famous works is ArnetMiner [3] [16], which offers comprehensive search and mining services for academic community. ArnetMiner not only provides abundant online academic services but also offers ideal test platform for heterogeneous information network analysis. PatentMiner [4] [17] is another application which is a general topic-driven framework for analyzing and mining heterogeneous patent networks.

8.4 Conclusions

With the surge of heterogeneous information network analysis, many prototype systems, even real systems, have been built based on heterogeneous networks. In this chapter, we introduce two prototype systems for recommendations. These prototype systems illustrate the advantages of heterogeneous information on semantics capture and information integration. However, compared to the boom of research on heterogeneous information network, the real applications are relatively insufficient. In the future, we need to solve practical problems in system construction, such as network construction with noise data, large-scale data processing, and scenario design of novel applications.

References

1. Balabanovic, M., Shoham, Y.: Content-based collaborative recommendation. Commun. ACM **40**(3), 66–72 (1997)
2. Breese, J., Heckerman, D., Kadie, C.: Empirical analysis of predictive algorithms for collaborative filtering. In: UAI, pp. 43–52 (1998)
3. Danilevsky, M., Wang, C., Tao, F., Nguyen, S., Chen, G., Desai, N., Wang, L., Han, J.: Amethyst: a system for mining and exploring topical hierarchies of heterogeneous data. In: KDD, pp. 1458–1461 (2013)
4. Han, J.: Mining heterogeneous information networks by exploring the power of links. In: DS, pp. 13–30 (2009)
5. Jin, X., Lin, C.X., Luo, J., Han, J.: Socialspamguard: a data mining-based spam detection system for social media networks. Proc. Vldb Endow. **4**(12), 1458–1461 (2011)
6. Jin, X., Wang, C., Luo, J., Yu, X., Han, J.: LikeMiner: a system for mining the power of 'like' in social media networks. In: KDD, pp. 753–756 (2011)
7. Lao, N., Cohen, W.: Fast query execution for retrieval models based on path constrained random walks. In: KDD, pp. 881–888 (2010)
8. Shang, M.S., Lu, L., Zhang, Y.C., Zhou, T.: Empirical analysis of web-based user-object bipartite networks. In: EPL, vol. 90(0120), p. 48006 (2010)
9. Sharma, A., Cosley, D.: Do social explanations work? Studying and modeling the effects of social explanations in recommender systems. In: WWW, pp. 1133–1143 (2013)
10. Shi, C., Kong, X., Yu, P.S., Xie, S., Wu, B.: Relevance search in heterogeneous networks. In: International Conference on Extending Database Technology, pp. 180–191 (2012)

---

[3] http://aminer.org/.

[4] http://pminer.org/home.do?m=home.

11. Shi, C., Li, Y., Zhang, J., Sun, Y., Yu, P.S.: A survey of heterogeneous information network analysis. Comput. Sci. **134**(12), 87–99 (2015)
12. Shi, C., Zhang, Z., Luo, P., Yu, P.S., Yue, Y., Wu, B.: Semantic path based personalized recommendation on weighted heterogeneous information networks. In: The ACM International, pp. 453–462 (2015)
13. Sun, Y., Han, J.: Mining heterogeneous information networks: a structural analysis approach. SIGKDD Explor. **14**(2), 20–28 (2012)
14. Sun, Y.Z., Han, J.W., Yan, X.F., Yu, P.S., Wu, T.: PathSim: meta path-based top-K similarity search in heterogeneous information networks. In: VLDB, pp. 992–1003 (2011)
15. Symeonidis, P., Nanopoulos, A., Manolopoulos, Y.: Moviexplain: a recommender system with explanations. In: RecSys, pp. 317–320 (2009)
16. Tang, J., Zhang, J., Yao, L., Li, J., Zhang, L., Su, Z.: ArnetMiner: extraction and mining of academic social networks. In: KDD, pp. 990–998 (2008)
17. Tang, J., Wang, B., Yang, Y., Hu, P., Zhao, Y., Yan, X., Gao, B., Huang, M., Xu, P., Li, W., Others: PatentMiner: topic-driven patent analysis and mining. In: KDD, pp. 1366–1374 (2012)
18. Tao, F., Yu, X., Lei, K.H., Brova, G., Cheng, X., Han, J., Kanade, R., Sun, Y., Wang, C., Wang, L., Others: Research-insight: providing insight on research by publication network analysis. In: SIGMOD, pp. 1093–1096 (2013)
19. Tao, F., Brova, G., Han, J., Ji, H., Wang, C., Norick, B., El-Kishky, A., Liu, J., Ren, X., Sun, Y.: NewsNetExplorer: automatic construction and exploration of news information networks. In: SIGMOD, pp. 1091–1094 (2014)
20. Tintarev, N., Masthoff, J.: A survey of explanations in recommender systems. In: ICDE Workshop, pp. 801–810 (2007)
21. Vig, J., Sen, S., Riedl, J.: Tagsplanations: explaining recommendations using tags. In: IUI, pp. 47–56 (2009)
22. Wang, B., Ester, M., Bu, J., Cai, D.: Who also likes it? Generating the most persuasive social explanations in recommender systems. In: AAAI, pp. 173–179 (2014)
23. Yu, X., Sun, Y., Zhao, P., Han, J.: Query-driven discovery of semantically similar substructures in heterogeneous networks. In: KDD, pp. 1500–1503 (2012)

Chapter 9
Future Research Directions

Abstract Although many data mining tasks have been exploited in heterogeneous information network, it is still a young and promising research field. Here, we illustrate some advanced topics, including challenging research issues and unexplored tasks, and point out some potential future research directions.

9.1 More Complex Network Construction

There is a basic assumption in contemporary researches that a heterogeneous information network to be investigated is well defined, and objects and links in the network are clean and unambiguous. However, it is not the case in real applications. In fact, constructing heterogeneous information network from real data often faces challenges.

If the networked data are structured data, like relational database, it may be easy to construct a heterogeneous information network with well-defined schema, such as DBLP network [36] and Movie network [28, 52]. However, even in this kind of heterogeneous networks, objects and links can still be noisy. (1) Objects in a network may not exactly correspond to entities in real world, such as duplication of name [47] in bibliography data. That is, one object in a network may refer to multiple entities, or different objects may refer to the same entity. We can integrate entity resolution [1] with network mining to clean objects or links beforehand. For example, Shen et al. [27] propose a probabilistic model SHINE to link named entity mentions detected from the unstructured Web text with their corresponding entities existing in a heterogeneous information network. Ren et al. [26] propose a relation phrase-based entity recognition framework, called ClusType. The framework runs data-driven phrase mining to generate entity mention candidates and relation phrases, and enforces the principle that relation phrases should be softly clustered when propagating type information in a heterogeneous network constructed by argument entities. (2) Relations among objects may not be explicitly given or not complete sometimes, e.g., the advisor–advisee relationship in the DBLP network [38]. Link prediction [18] can be employed to fill out the missing relations for comprehensive networks. (3) Objects and links may not be reliable or trustable, e.g., the inaccurate item information in an

© Springer International Publishing AG 2017

C. Shi and P.S. Yu, *Heterogeneous Information Network Analysis and Applications*, Data Analytics, DOI 10.1007/978-3-319-56212-4_9

E-commerce Web site and conflicting information of certain objects from multiple Web sites. However, an HIN can be built to capture the dependency relations among the node entities to clean up and integrate the data, such as trustworthiness modeling [48, 59], spam detection [45], and co-ranking of questions, answers, and users in a Q&A system.

If the networked data are unstructured data, such as text data, multimedia data, and multilingual data, it becomes more challenging to construct qualified heterogeneous information networks. In order to construct high-quality HINs, information extraction, natural language processing, and many other techniques should be integrated with network construction. Mining quality phrases is a critical step to form entities of networks from text data. Kishky et al. [6] propose a computationally efficient and effective model ToPMine, which first executes a phrase mining framework to segment a document into single and multiword phrases, and then employs a new topic model that operates on the induced document partition. Furthermore, Liu et al. [21] propose an effective and scalable method SegPhrase+ that integrates quality phrases extraction with phrasal segmentation. Beyond the bag-of-word representation of text data, some researchers try to represent a document with the help of heterogeneous information network. Wang et al. [41] firstly map entities in documents into a knowledge base (e.g., Freebase), and then consider the knowledge base as an HIN to mine internal relations among entities. Furthermore, Wang et al. [40, 43] employ world knowledge as indirect supervision to improve the document clustering results. More recently, Wang et al. [42] propose the HIN-kernel concept for classification through representing a text as an HIN. Relationship extraction is another important step to form links among the objects in network. Wang et al. [38] mine hidden advisor–advisee relationships from bibliographic data, and they further infer hierarchical relationships among partially ordered objects with heterogeneous attributes and links [39]. Broadly speaking, we can also extract entity and relationship to construct heterogeneous network from multimedia data and multilingual data, as we have done on text data.

9.2 More Powerful Mining Methods

For ubiquitous heterogeneous information networks, numbers of mining methods have been proposed on many data mining tasks. As we have mentioned, heterogeneous information networks have two important characteristics: complex structure and rich semantics. According to these two characteristics, we summarize the contemporary works and point out future directions.

9.2.1 Network Structure

In heterogeneous network, objects can be organized in different forms. Bipartite graph is widely used to organize two types of objects and the relations between them [10, 23]. As an extension of bipartite graphs, K-partite graphs [22] are able to represent multiple types of objects. Recently, heterogeneous networks are usually organized as star-schema networks, such as bibliographic data [29, 34, 36] and movie data [28, 52]. To combine the heterogeneous and homogeneous information, star-schema with self loop is also proposed [46]. Different from only one hub object type existing in star-schema network, some networked data have multiple hub object types, e.g., the bioinformatics data [31]. For this kind of networks, Shi et al. [31] propose a HeProjI method which projects a general heterogeneous network into a sequence of subnetworks with bipartite or star-schema structure.

In applications, the networked data are usually more complex and irregular. Some real networks may contain attribute values on links, and these attribute values may contain important information. For example, users usually rate movies with a score from 1 to 5 in movie recommended system, where the rating scores represent users' attitudes to movies, and the "author of" relation between authors and papers in bibliographic networks can take values (e.g., 1, 2, 3) which represents the order of authors in the paper. In this kind of applications, we need to consider the effect of attribute values on the weighted heterogeneous information network [32]. There are some time series data, for example, a period of biographic data and rating information of users and movies. For this kind of data, we need to construct dynamic heterogeneous network [35] and consider the effect of the time factor. In some applications, one kind of objects may exist in multiple heterogeneous networks [12, 54]. For example, users usually co-exist in multiple social networks, such as Facebook, Google+, and Twitter. In this kind of applications, we need to align users in different networks and effectively fuse information from different networks [55–57]. More broadly, many networked data are difficult to be modeled with heterogeneous network with a simple network schema. For example, in RDF data, there are so many types of objects and relations, which cannot be described with network schema [25, 40]. Many research problems arise with this kind of schema-rich HINs [3, 44], for example, management of objects and relations with so many types and automatic generation of meta paths. As the real networked data become more complex, we need to design more powerful and flexible heterogeneous networks, which also provides more challenges for data mining.

9.2.2 Semantic Mining

As the unique characteristic, objects and links in HIN contain rich semantics. Meta path can effectively capture subtle semantics among objects, and many works have made use of the meta path-based mining tasks. For example, in similarity measure task, object pairs have different similarities under different meta paths [29, 36]; in recommendation task, different items will be recommended under different paths [32]. In addition, meta path is also widely used for feature extraction. Object similarity can be measured under different meta paths, which can be used as feature vectors for many tasks, such as clustering [37], link prediction [2], and recommendation [53].

However, some researchers have noticed the shortcomings of meta path. Since meta path fails to capture more microsemantics. In some applications, some researchers consider to refine meta path with some constraints. For example, the "Author-Paper-Author" path describes the collaboration relation among authors. However, it cannot depict the fact that Philip S. Yu and Jiawei Han have many collaborations in data mining field but they seldom collaborate in information retrieval field. In order to overcome the shortcoming existing in meta path, Shi et al. [16] propose the constrained meta path concept, which can confine some constraints on objects. Taking Fig. 1.3c in Chap. 1 as an example, the constrained meta path $APA|P.L =$ "$Data\ Mining$" represents the co-author relation of authors in data mining field through constraining the label of papers with "Data Mining." Moreover, Liu et al. [20] propose the concept "restricted meta-path" which enables in-depth knowledge mining on the heterogeneous bibliographic networks by allowing restrictions on the node set. In addition, traditional HIN and meta path do not consider the attribute values on links, while weighted links are very common in practical applications. Examples include rating scores between users and items in recommended system and the order of authors in papers in bibliographic network. Taking Fig. 5.2 in Chap. 5 as an example, the rating relation between users and movies can take scores from 1 to 5. Shi et al. [32] propose weighted meta path to consider attribute values on links and more subtly capture path semantics through distinguishing different link attribute values.

On the other hand, some researchers consider to capture more macro semantics through combining multiple-related meta paths. For example, two authors write two different papers that both mention the mining term and are published in the same venue, while another two authors also write two different papers that are published in the same venue and have not the same terms. Therefore, these two authors in the first case should have a higher relevance score than those two authors in the second case. However, the single meta path either $APVPA$ or $APTPA$ fails to discover this factor. In order to solve this shortcoming, Huang et al. [9] propose the relevance measure based on metastructure which is a combination of meta paths. Similarly, Fang et al. [7] propose the metagraph which is a subgraph defined on a graph schema and can measure the semantic proximity between objects. As an effective semantic capture tool, meta path has shown its power in semantic capture and feature selection. However, it may be coarse in some applications, so we need to extend traditional meta path for more subtle semantic capture. Broadly speaking, we can also design new and more powerful semantic capture tools.

More importantly, the meta path approach faces challenges on path selection and their weight importances. How can we select meta paths in real applications? Theoretically, there are infinite meta paths in an HIN. In contemporary works, the network schema of HIN is usually small and simple, so we can assign some short

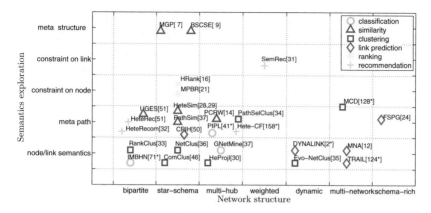

Fig. 9.1 Summarization of typical works on HIN according to network structure and semantic exploration. '∗' in a citation means this citation is from Chap. 2

and meaningful meta paths according to domain knowledge and experiences. Sun et al. [36] have validated that the long meta paths are not meaningful and they fail to produce good similarity measures. However, there is no work to study the effect of long meta paths on other mining tasks. In addition, there are so many meta paths even for short paths in some complex networks, like RDF network. It is a critical task to extract meta paths automatically in this condition. Recently, Meng et al. [25] study how to discover meta paths automatically which can best explain the relationship between node pairs. Another important issue is to determine the weights of meta paths automatically. Some methods have been proposed to explore this issue. For example, Lao et al. [14] employ a supervised method to learn weights, and Sun et al. [37] combine meta-path selection and user-guided information for clustering. In addition, Liang et al. [17] seek to find the K, most interesting path instances matching the preferred relationship type. Some interesting works are still worth doing. The ideal path weights learned should embody the importance of paths and reflect users' preferences. However, the similarity evaluations based on different paths have significant bias, which may make path weights hard to reflect path importances. So prioritized path weights are needed. In addition, if there are numerous meta paths in real applications (e.g., RDF network), the path weight learning will be more important and challenging.

In Fig. 9.1, we summarize some typical works in the HIN field from two perspectives: network structure and semantic exploration. We respectively select several typical works from six mining tasks mentioned above and put these works in a coordinate according to network structure and semantics exploration in these works. Note that we denominate those un-named methods with the first letter of keywords in the title, such as UGES [51] and CPIH [50]. Along the X-axis, the network structure becomes more complex, and semantics information becomes richer along the Y-axis. For example, RankClus [33] is designed for bi-type networks and only captures link semantics (different-typed links contain different semantics), while PathSim [36] can deal with more complex star-schema networks and use meta path to capture deeper semantics. Further, SemRec [32] adds constraints to links to explore more subtle semantic information in a weighted HIN. From the figure, we can also find that most contemporary works focus on simple network structures (e.g., bipartite or star-schema networks) and primary semantic exploration (e.g., meta path). In the future, we can exploit more complex heterogeneous networks with more powerful semantics capture tools.

9.3 Bigger Networked Data

In order to illustrate the benefits of HIN, we need to design data mining algorithms on big-networked data in wider domains. This variety is an important characteristic of big data. HIN is a powerful tool to handle the diversity of big data, since it can flexibly and effectively integrate varied objects and heterogeneous information. However, it is non-trivial work to build a real HIN-based analysis system. Besides research challenges mentioned above, such as network construction, it will face many practical technique challenges. A real HIN is huge, even dynamic, so it usually cannot be contained in memory and cannot be handled directly. We know that a user at a time could be only interested in a tiny portion of nodes, links, or subnetworks. Instead of directly mining the whole network, we can mine hidden but small networks "extracted" dynamically from some existing networks, based on user-specified constraints or expected node/link behaviors. How to discover such hidden networks and mine knowledge (e.g., clusters, behaviors, and anomalies) from such hidden but non-isolated networks could be an interesting but challenging problem.

Most of contemporary data mining tasks on HIN only work on small dataset and fail to consider the quick and parallel process on big data. Some research works have begun to consider the quick computation of mining algorithms on HIN. For example, Sun et al. [36] design a co-clustering-based pruning strategy to fasten the processing speed of PathSim. Lao et al. [13] propose the quick computation strategies of PCRW, and Shi et al. [24, 30] also consider the quick/parallel computation of HeteSim. In addition, cloud computing also provides an option to handle big-networked data. Although parallel graph mining algorithms [4] and platforms [11] have been proposed, parallel HIN analysis methods face some unique challenges. For example, the partition of HIN needs to consider the overload balances of computing nodes, as well as balances of different-typed nodes. Moreover, it is also challenging to mine integrated path semantics in partitioned subgraphs.

9.4 More Applications

Due to unique characteristics of HIN, many data mining tasks have been explored on HIN, which are summarized as above. In fact, more data mining tasks can be studied on HIN. Here, we introduce two potential applications.

The online analytical processing (OLAP) has shown its power in multidimensional analysis of structured relational data [5]. The similar analysis can also be done, when we view a heterogeneous information network from different angles and at different levels of granularity. Taking a bibliographic network as an example, we can observe the change of published papers on a conference in the time or district dimension, when we designate papers and conferences as the object types and publish relations as the link type. Some preliminary studies have been done on this issue. Zhao et al. [58] introduce graph cube to support OLAP queries effectively on large multidimensional networks; Li et al. [15] design InfoNetOLAPer to provide topic-oriented, integrated, and multidimensional organizational solutions for information networks. Yin et al. [49] have developed a novel HMGraph OLAP framework to mine multidimensional heterogeneous information networks with more dimensions and operations. These works consider link relation as a measure. However, they usually ignore semantic information in heterogeneous networks determined by multiple nodes and links. So the study of online analytical processing of heterogeneous information networks is still worth exploring.

Information diffusion is a vast research domain and has attracted research interests from many fields, such as physics and biology. Traditional information diffusion is studied on homogeneous networks [8], where information is propagated in one single channel. However, in many applications, pieces of information or diseases are propagated among different types of objects. For example, diseases could propagate among people, different kinds of animals, and food, via different channels. Few works explore this issue. Liu et al. [19] propose a generative graphical model which utilizes the heterogeneous link information and the textual content associated with each node to mine topic-level direct influence. In order to capture better spreading models that represent the real-world patterns, it is desirable to pay more attention to the study of information diffusion in heterogeneous information networks.

References

1. Bhattacharya, I., Getoor, L.: Collective entity resolution in relational data. ACM Trans. Knowl. Discov. Data **1**(1), 5 (2007)
2. Cao, B., Kong, X., Yu, P.S.: Collective prediction of multiple types of links in heterogeneous information networks. In: ICDM, pp. 50–59 (2014)
3. Cao, X., Zheng, Y., Shi, C., Li, J., Wu, B.: Link prediction in Schema-Rich heterogeneous information network. In: PAKDD, pp. 449–460 (2016)
4. Cohen, J.: Graph twiddling in a MapReduce world. Comput. Sci. Eng. **11**(4), 29–41 (2009)
5. Colliat, G.: OLAP, relational, and multidimensional database systems. ACM Sigmod Rec. **25**(3), 64–69 (1996)
6. El-Kishky, A., Song, Y., Wang, C., Voss, C.R., Han, J.: Scalable topical phrase mining from text corpora. PVLDB **8**(3), 305–316 (2014)
7. Fang, Y., Lin, W., Zheng, V.W., Wu, M., Chang, C.C., Li, X.L.: Semantic proximity search on graphs with metagraph-based learning. In: ICDE, pp. 277–288 (2016)
8. Gruhl, D., Guha, R., Liben-Nowell, D., Tomkins, A.: Information diffusion through blogspace. In: WWW, pp. 491–501 (2004)

9. Huang, Z., Zheng, Y., Cheng, R., Sun, Y., Mamoulis, N., Li, X.: Meta structure: Computing relevance in large heterogeneous information networks. In: KDD, pp. 1595–1604 (2016)

10. Jamali, M., Lakshmanan, L.V.S.: HeteroMF: recommendation in heterogeneous information networks using context dependent factor models. In: WWW, pp. 643–654 (2013)

11. Kang, U., Tsourakakis, C.E., Faloutsos, C.: Pegasus: A peta-scale graph mining system implementation and observations. In: ICDM, pp. 229–238 (2009)

12. Kong, X., Zhang, J., Yu, P.S.: Inferring anchor links across multiple heterogeneous social networks. In: CIKM, pp. 179–188 (2013)

13. Lao, N., Cohen, W.: Fast query execution for retrieval models based on path constrained random walks. In: KDD, pp. 881–888 (2010)

14. Lao, N., Cohen, W.W.: Relational retrieval using a combination of path-constrained random walks. Mach. Learn. **81**(2), 53–67 (2010)

15. Li, C., Yu, P.S., Zhao, L., Xie, Y., Lin, W.: InfoNetOLAPer: integrating InfoNetWarehouse and InfoNetCube with InfoNetOLAP. PVLDB **4**(12), 1422–1425 (2011)

16. Li, Y., Shi, C., Yu, P.S., Chen, Q.: HRank: a path based ranking method in heterogeneous information network. In: WAIM, pp. 553–565 (2014)

17. Liang, J., Ajwani, D., Nicholson, P.K., Sala, A., Parthasarathy, S.: What links alice and bob?: Matching and ranking semantic patterns in heterogeneous networks. In: WWW, pp. 879–889 (2016)

18. Liben-Nowell, D., Kleinberg, J.: The link-prediction problem for social networks. J. Am. Soc. Inform. Sci. Technol. **58**(7), 1019–1031 (2007)

19. Liu, L., Tang, J., Han, J., Jiang, M., Yang, S.: Mining topic-level influence in heterogeneous networks. In: CIKM, pp. 199–208 (2010)

20. Liu, X., Yu, Y., Guo, C., Sun, Y.: Meta-path-based ranking with pseudo relevance feedback on heterogeneous graph for citation recommendation. In: CIKM, pp. 121–130 (2014)

21. Liu, J., Shang, J., Wang, C., Ren, X., Han, J.: Mining quality phrases from massive text corpora. In: SIGMOD, pp. 1729–1744 (2015)

22. Long, B., Wu, X., Zhang, Z., Yu, P.S.: Unsupervised learning on k-partite graphs. In: KDD, pp. 317–326 (2006)

23. Long, B., Zhang, Z.M., Yu, P.S.: Co-clustering by block value decomposition. In: KDD, pp. 635–640 (2005)

24. Meng, X., Shi, C., Li, Y., Zhang, L., Wu, B.: Relevance measure in large-scale heterogeneous networks. In: APWeb, pp. 636–643 (2014)

25. Meng, C., Cheng, R., Maniu, S., Senellart, P., Zhang, W.: Discovering meta-paths in large heterogeneous information networks. In: WWW, pp. 754–764 (2015)

26. Ren, X., El-Kishky, A., Wang, C., Tao, F., Voss, C.R., Han, J.: ClusType: effective entity recognition and typing by relation phrase-based clustering. In: KDD, pp. 995–1004 (2015)

27. Shen, W., Han, J., Wang, J.: A probabilistic model for linking named entities in web text with heterogeneous information networks. In: SIGMOD, pp. 1199–1210 (2014)

28. Shi, C., Zhou, C., Kong, X., Yu, P.S., Liu, G., Wang, B.: HeteRecom: a semantic-based recommendation system in heterogeneous networks. In: KDD, pp. 1552–1555 (2012)

29. Shi, C., Kong, X., Yu, P.S., Xie, S., Wu, B.: Relevance search in heterogeneous networks. In: International Conference on Extending Database Technology, pp. 180–191 (2012)

30. Shi, C., Kong, X., Huang, Y., Philip, S.Y., Wu, B.: Hetesim: a general framework for relevance measure in heterogeneous networks. IEEE Trans. Knowl. Data Eng. **26**(10), 2479–2492 (2014)

31. Shi, C., Wang, R., Li, Y., Yu, P.S., Wu, B.: Ranking-based clustering on general heterogeneous information networks by network projection. In: CIKM, pp. 699–708 (2014)

32. Shi, C., Zhang, Z., Luo, P., Yu, P.S., Yue, Y., Wu, B.: Semantic path based personalized recommendation on weighted heterogeneous information networks. In: CIKM, pp. 453–462 (2015)

33. Sun, Y., Han, J., Zhao, P., Yin, Z., Cheng, H., Wu, T.: RankClus: Integrating clustering with ranking for heterogeneous information network analysis. In: EDBT, pp. 565–576 (2009)

34. Sun, Y., Yu, Y., Han, J.: Ranking-based clustering of heterogeneous information networks with star network schema. In: KDD, pp. 797–806 (2009)

35. Sun, Y., Tang, J., Han, J., Gupta, M., Zhao, B.: Community evolution detection in dynamic heterogeneous information networks. In: MLG, pp. 137–146 (2010)
36. Sun, Y.Z., Han, J.W., Yan, X.F., Yu, P.S., Wu, T.: PathSim: meta path-based Top-K similarity search in heterogeneous information networks. In: VLDB, pp. 992–1003 (2011)
37. Sun, Y., Norick, B., Han, J., Yan, X., Yu, P.S., Yu, X.: Integrating meta-path selection with user-guided object clustering in heterogeneous information networks. In: KDD, pp. 1348–1356 (2012)
38. Wang, C., Han, J., Jia, Y., Tang, J., Zhang, D., Yu, Y., Guo, J.: Mining advisor-advisee relationships from research publication networks. In: KDD, pp. 203–212 (2010)
39. Wang, C., Han, J., Li, Q., Li, X., Lin, W.P., Ji, H.: Learning hierarchical relationships among partially ordered objects with heterogeneous attributes and links. In: SDM, pp. 516–527 (2012)
40. Wang, C., Song, Y., El-Kishky, A., Roth, D., Zhang, M., Han, J.: Incorporating world knowledge to document clustering via heterogeneous information networks. In: KDD, pp. 1215–1224 (2015)
41. Wang, C., Song, Y., Li, H., Zhang, M., Han, J.: Knowsim: A document similarity measure on structured heterogeneous information networks. In: ICDM, pp. 1015–1020 (2015)
42. Wang, C., Song, Y., Li, H., Zhang, M., Han, J.: Text classification with heterogeneous information network kernels. In: AAAI, pp. 2130–2136 (2016)
43. Wang, C., Song, Y., Roth, D., Zhang, M., Han, J.: World knowledge as indirect supervision for document clustering (2016). arXiv preprint. arXiv:1608.00104
44. Wang, C., Sun, Y., Song, Y., Han, J., Song, Y., Wang, L., Zhang, M.: Relsim: relation similarity search in schema-rich heterogeneous information networks. In: Siam International Conference on Data Mining, pp. 621–629 (2016)
45. Wang, G., Xie, S., Liu, B., Yu, P.S.: Identify online store review spammers via social review graph. TIST 3(4), 61 (2012)
46. Wang, R., Shi, C., Yu, P.S., Wu, B.: Integrating clustering and ranking on hybrid heterogeneous information network. In: PAKDD, pp. 583–594 (2013)
47. Yin, X., Han, J., Yu, P.S.: Object distinction: distinguishing objects with identical names. In: ICDE, pp. 1242–1246 (2007)
48. Yin, X., Han, J., Yu, P.S.: Truth discovery with multiple conflicting information providers on the web. Knowl. Data Eng. 20(6), 796–808 (2008)
49. Yin, M., Wu, B., Zeng, Z.: HMGraph OLAP: a novel framework for multi-dimensional heterogeneous network analysis. In: DOLAP, pp. 137–144 (2012)
50. Yu, X., Gu, Q., Zhou, M., Han, J.: Citation prediction in heterogeneous bibliographic networks. In: SDM, pp. 1119–1130 (2012)
51. Yu, X., Sun, Y., Norick, B., Mao, T., Han, J.: User guided entity similarity search using meta-path selection in heterogeneous information networks. In: CIKM, pp. 2025–2029 (2012)
52. Yu, X., Ren, X., Sun, Y., Sturt, B., Khandelwal, U., Gu, Q., Norick, B., Han, J.: Recommendation in heterogeneous information networks with implicit user feedback. In: RecSys, pp. 347–350 (2013)
53. Yu, X., Ren, X., Sun, Y., Gu, Q., Sturt, B., Khandelwal, U., Norick, B., Han, J.: Personalized entity recommendation: a heterogeneous information network approach. In: WSDM, pp. 283–292 (2014)
54. Zhang, J., Kong, X., Yu, P.S.: Predicting social links for new users across aligned heterogeneous social networks. In: ICDM, pp. 1289–1294 (2013)
55. Zhang, J., Shao, W., Wang, S., Kong, X., Yu, P.S.: Partial network alignment with anchor meta path and truncated generic stable matching. ArXiv e-prints (2015)
56. Zhang, J., Yu, P.S.: Integrated anchor and social link predictions across social networks. In: IJCAI, pp. 2125–2131 (2015)
57. Zhang, J., Yu, P.S.: Multiple anonymized social networks alignment. In: ICDM, pp. 599–608 (2015)
58. Zhao, P., Li, X., Xin, D., Han, J.: Graph cube: on warehousing and OLAP multidimensional networks. In: SIGMOD, pp. 853–864 (2011)
59. Zhao, B., Rubinstein, B.I.P., Gemmell, J., Han, J.: A bayesian approach to discovering truth from conflicting sources for data integration. PVLDB 5(6), 550–561 (2012)

Printed in the United States
By Bookmasters